T0350997

What Every Engineer Should Know about Software Engineering

This book offers a practical approach to understanding, designing, and building sound software based on solid principles. Using a unique Q&A format, this book addresses the issues that engineers need to understand in order to successfully work with software engineers, develop specifications for quality software, and learn the basics of the most common programming languages, development approaches, and paradigms. The new edition is thoroughly updated to improve the pedagogical flow and emphasize new software engineering processes, practices, and tools that have emerged in every software engineering area.

Features:

- Defines concepts and processes of software and software development, such as agile processes, requirements engineering, and software architecture, design, and construction.
- Uncovers and answers various misconceptions about the software development process and presents an up-to-date reflection on the state of practice in the industry.
- Details how non-software engineers can better communicate their needs to software engineers and more effectively participate in design and testing to ultimately lower software development and maintenance costs.
- Helps answer the question: How can I better leverage embedded software in my design?
- Adds new chapters and sections on software architecture, software engineering and systems, and software engineering and disruptive technologies, as well as information on cybersecurity.
- Features new appendices that describe a sample automation system, covering software requirements, architecture, and design.

This book is aimed at a wide range of engineers across many disciplines who work with software.

What Every Engineer Should Know
Series Editor
Phillip A. Laplante
Pennsylvania State University

What Every Engineer Should Know about Developing Real-Time Embedded Products
Kim R. Fowler

What Every Engineer Should Know about Business Communication
John X. Wang

What Every Engineer Should Know about Career Management
Mike Ficco

What Every Engineer Should Know about Starting a High-Tech Business Venture
Eric Koester

What Every Engineer Should Know about MATLAB® and Simulink®
Adrian B. Biran

Green Entrepreneur Handbook: The Guide to Building and Growing a Green and Clean Business
Eric Koester

What Every Engineer Should Know about Cyber Security and Digital Forensics
Joanna F. DeFranco

What Every Engineer Should Know about Modeling and Simulation
Raymond J. Madachy and Daniel Houston

What Every Engineer Should Know about Excel, Second Edition
J.P. Holman and Blake K. Holman

Technical Writing: A Practical Guide for Engineers, Scientists, and Nontechnical Professionals, Second Edition
Phillip A. Laplante

What Every Engineer Should Know about the Internet of Things
Joanna F. DeFranco and Mohamad Kassab

What Every Engineer Should Know about Software Engineering
Phillip A. Laplante and Mohamad Kassab

What Every Engineer Should Know about Cyber Security and Digital Forensics
Joanna F. DeFranco and Bob Maley

For more information about this series, please visit: www.routledge.com/
What-Every-Engineer-Should-Know/book-series/CRCWEESK

What Every Engineer Should Know about Software Engineering

Second Edition

Phillip A. Laplante and Mohamad Kassab

CRC Press
Taylor & Francis Group
Boca Raton London New York

CRC Press is an imprint of the
Taylor & Francis Group, an **informa** business

Second edition published 2023
by CRC Press
6000 Broken Sound Parkway NW, Suite 300, Boca Raton, FL 33487-2742

and by CRC Press
4 Park Square, Milton Park, Abingdon, Oxon, OX14 4RN

CRC Press is an imprint of Taylor & Francis Group, LLC

© 2023 Phillip A. Laplante and Mohamad Kassab

First edition published by CRC Press 2007

ISBN: 978-1-032-10318-1 (hbk)
ISBN: 978-1-032-11153-7 (pbk)
ISBN: 978-1-003-21864-7 (ebk)

DOI: 10.1201/9781003218647

Typeset in Times
by Newgen Publishing UK

Phillip A. Laplante dedicates this book to the countless people—practitioners, researchers and students—who have helped him better understand software engineering over the last 40 years.

Mohamad Kassab dedicates this book to his family members; Hassan, Mariam, Essa, Eman, Sana, and Samar for the support, and love they provided during the journey of writing this edition.

Contents

Authors

Phillip A. Laplante is Professor of Software and Systems Engineering at the Pennsylvania State University. He earned a BS, an MEng, and a PhD in computer science at Stevens Institute of Technology and an MBA at the University of Colorado. He is a Fellow of the IEEE and SPIE and has won international awards for his teaching, research, and service. From 2010 to 2017 he led the effort to develop a national licensing exam for software engineers.

He has worked in avionics, CAD, and software testing systems and he has published 37 books and more than 250 scholarly papers. He is a licensed professional engineer in the Commonwealth of Pennsylvania. He is also a frequent technology advisor to senior executives, investors, entrepreneurs, and attorneys and actively serves on corporate technology advisory boards.

His research interests include artificial intelligent systems, critical systems, requirements engineering and software quality and management. Prior to his appointment at Penn State, he was a software development professional, technology executive, college president, and entrepreneur.

Over the years he has worked with, and for, many kinds of engineers. Non-software engineers have worked with him as well, and he has had the pleasure of teaching thousands of practicing engineers of various types about software engineering. This text, then, represents a compendium of what engineers should know about software engineering. He also consults regularly for the software industry, including Fortune 1000 companies and smaller software development houses.

Mohamad Kassab is Associate Research Professor and a member of the graduate faculty at the Pennsylvania State University. He earned an MS and a PhD in computer science at Concordia University in Montreal, Canada. Dr. Kassab was an affiliate assistant professor in the Department of Computer Science and Software Engineering at Concordia University between 2010 and 2012 and a postdoctoral researcher in software engineering at Ecole de Technologie Supérieure (ETS) in Montreal between 2011 and 2012 and a visiting scholar at Carnegie Mellon University (CMU) between 2014 and 2015.

Dr. Kassab has been conducting research projects jointly with the industry to develop formal and quantitative models to support the integration of quality requirements within software and systems development life cycles. The models are being further leveraged with the support of developed architectural frameworks and tools. His research interests also include bridging the gap between software engineering practices and disruptive technologies (e.g., IoT, blockchain). He has published extensively in software engineering books, journals, and conference proceedings. He is also a member of numerous professional societies and program committees, and the organizer of many software engineering workshops and conference sessions.

With over twenty years of global industry experience, Dr. Kassab has developed a broad spectrum of skills and responsibilities in many software engineering areas.

Notable experiences include business unit manager at Soramitsu, senior quality engineer at SAP, senior quality engineer at McKesson, senior associate at Morgan Stanley, senior quality assurance specialist at NOKIA, and senior software developer at Positron Safety Systems. He is an Oracle Certified Application Developer, Sun Certified Java Programmer, and Microsoft Certified Professional.

Dr. Kassab has taught a variety of graduate and undergraduate software engineering and computer science courses at Penn State and Concordia University. He has won many awards for his excellence in teaching.

Introduction

WHAT IS THE GOAL OF THIS BOOK?

This is a book about software engineering, but its purpose is not to enable you to leap into the role of a fully trained software engineer. That goal would be impossible to achieve solely with the reading of any book. Instead, the goal of this book is to help you better understand the nature of software engineering as a profession, as an engineering discipline, as a culture, and as an art form. And it is because of its ever morphing, multidimensional nature that non-software engineers have so much difficulty understanding the challenges software engineers must face.

Many practicing software engineers have little or no formal education in software engineering. While software engineering is a discipline in which practice and experience are important, it is rare that someone who has not studied software engineering will have the skills and knowledge needed to efficiently build industrial strength, reliable software systems. While these individuals may be perfectly capable of building working systems, unless a deliberate software engineering approach is followed, the cost of development will probably be higher than necessary, and the cost of maintaining the system will certainly be higher than it ought to be.

HOW IS THIS BOOK DIFFERENT FROM OTHER SOFTWARE ENGINEERING BOOKS?

It is different from other software engineering books and from every other conventional engineering textbook in that is written using the Socratic Method; that is, in the form of questions and answers. In some places we have reused material from other books written by the authors, with permission, particularly *Software Engineering for Image Processing Systems*; *Requirements Engineering for Software and Systems, Fourth Edition;* and *Antipatterns: Identification, Refactoring, and Management, Second Edition*. But because of the Socratic format, significant rewriting was needed to place the material in the appropriate form of discourse. Furthermore, in this present text we have generalized the concepts from that of predecessor texts to address the broader needs of all kinds of engineers.

CAN THIS BOOK CONVERT ME INTO A SOFTWARE ENGINEER?

We don't promise that after reading this book you will become a master software engineer—no book can deliver on that promise. What this book will do, we hope, is help you better understand the limits of software engineering and the advances that have been made over the years. If nothing else, it is our hope that you will come away

DOI: 10.1201/9781003218647-1

from reading this book with a deeper, more sympathetic understanding of the software engineer as an engineer.

ARE SOFTWARE ENGINEERS REALLY ENGINEERS?

Yes, the software engineer can be regarded as an engineer if they have the proper education, discipline, experience, and mindset.

HOW SHOULD I USE THIS BOOK?

To get the most benefit from this book, we suggest you use it in one or more of the following ways:

- Read it through in its entirety to provide a general framework for and understanding of the profession of software engineering.
- Use it as a regular reference when questions about software, software engineering, or software engineers arise. You will find most of your questions directly addressed in this book.
- Skip around and read sections as needed to answer specific questions. There is no harm in reading this book out of order; after all, it was written out of order.

WHO IS THE INTENDED AUDIENCE?

The target reader is the practicing engineer who has found they must write software, integrate off-the-shelf software into the systems they build, or who works with software engineers on a regular basis. Undergraduate and graduate engineering students would be well served to have this book for reference, as it is likely that they will find themselves in the position of building software, and it is good to establish a rigorous framework early in their careers.

DID ANYONE HELP YOU WITH THIS BOOK?

Many people have contributed bits and pieces to predecessors of this text or motivated our writing in discussions. Over the years, many students in our graduate software engineering courses (often with degrees in various engineering disciplines) have contributed ideas that have influenced this book.

There are so many people that we can't thank them individually, only collectively.

We must, however, thank the wonderful folks at CRC Press/Taylor & Francis, particularly our senior editor, Allison Shatkin, and the Editorial Director of Engineering, Nora Konopka.

ARE THERE COPYRIGHTS AND TRADEMARKS TO BE CITED?

All companies are the holders of the respective trademarks for any products mentioned in this text.

As noted previously, some of this book has been excerpted or adapted, with permission from the authors' own texts or others that are published by the CRC Press/Taylor & Francis Publishing Group. These are:

- Phillip A. Laplante and Mohamad H. Kassab, *Requirements Engineering for Software and Systems*, Fourth Edition, Taylor & Francis, 2022.
- *Antipatterns: Identification, Refactoring, and Management*, Second Edition, Colin J. Neill, Phillip A. Laplante, and Joanna DeFranco, Taylor & Francis, 2012.
- *Dictionary of Computer Science, Engineering, and Technology*, Phillip A. Laplante (Editor), CRC Press, 2001.
- *Lightweight Enterprise Architectures*, Fenix Theuerkorn, Auerbach Publications, 2005.
- *Real Process Improvement Using CMMI*, Michael West, Auerbach Publications, 2004.
- *Software Engineering for Image Processing Systems*, Phillip A. Laplante, CRC Press, 2003.
- *Software Engineering Handbook*, Jessica Keyes, Auerbach Publications, 2003.
- *Software Engineering Measurement*, John C. Munson, Auerbach Publications, 2003.
- *Software Engineering Processes: Principles and Applications*, Yingxu Wang and Graham King, CRC Press, 2000.
- *Software Engineering Quality Practices*, Kurt Kandt, Auerbach Publications, 2005.
- *Software Testing and Continuous Quality Improvement*, Second Edition, William E. Lewis, Auerbach Publications, 2005.
- *Software Testing: A Craftsman's Approach*, Second Edition, Paul Jorgensen, CRC Press, 2002.
- *The Computer Science and Engineering Handbook*, Allen B. Tucker, Jr. (editor-in-chief), CRC Press, 1997.

Where more substantial portions of material have been used verbatim, or figures reproduced, it is so noted. Otherwise, these texts may be considered part of the general reference material for preparing this book.

1 The Profession of Software Engineering

OUTLINE

- Software engineering as an engineering profession
- Standards and certifications
- Misconceptions about software engineering

1.1 INTRODUCTION

If you want to start a debate among your engineering friends, ask the question, "Is software engineering real engineering?" Unfortunately, we suspect that if your friends are from one of the "hard" engineering disciplines such as mechanical, civil, chemical, and electrical, then their answers may be "no." This is unfortunate because software engineers have been trying for many years to elevate their profession to a level of respect granted to the hard engineering disciplines. There are strong feelings around many aspects of the practice of software engineering—licensure, standards, minimum education, and so forth.

Therefore, it is appropriate to start a book about software engineering by focusing on these fundamental issues.

1.2 SOFTWARE ENGINEERING AS AN ENGINEERING PROFESSION

1.2.1 WHAT IS SOFTWARE ENGINEERING?

Software engineering is "a systematic approach to the analysis, design, assessment, implementation, test, maintenance and reengineering of software, that is, the application of engineering to software. In the software engineering approach, several models for the software life cycle are defined, and many methodologies for the definition and assessment of the different phases of a life-cycle model" (Laplante, 2001).

The profession of software engineering encompasses all aspects of conceiving, communicating, specifying, designing, building, testing, and maintaining software systems. Software engineering activities also include everything to do with the production of the artifacts related to software engineering such as documentation and tools.

There are many other ancillary activities to software engineering, one of which is the programming of the code or coding. But if you were stopped on the street by a pedestrian and asked to give a one-word definition for software engineering, your answer should be, "modeling." If you had two words to give, you might say, "modeling" and "optimization."

DOI: 10.1201/9781003218647-2

Modeling is a translation activity. The software product concept is translated into a requirements specification. The requirements are converted into architecture and design. The design is then converted into code, which is automatically translated by compilers and assemblers, which produce machine-executable code. In each of these translation steps, however, errors are likely to be introduced either by the humans involved or by the tools they use. Thus, the practice of software engineering involves reducing translation errors through the application of correct principles.

The optimization part deals with finding the most economical translation possible. "Economical" means in terms of efficiency, clarity, and other desirable properties, which will be discussed later.

1.2.2 Is Software Engineering an Engineering Discipline?

The answer to this question depends on who you ask. Many readers will argue that software engineering is not a true engineering discipline because there are no fundamental theorems grounded in the laws of physics (more on this later). Even some software engineering experts would add that there is still too much "art" in software engineering; that is, ad hoc approaches instead of rigorous ones. To further tarnish the image of software engineering, many self-styled practitioners do not have the appropriate background to engage in software engineering. These frauds help propagate the worst stereotypes by exemplifying what software engineering is not and should not be.

Perhaps the greatest assault on the reputation of software engineering and engineers occurs because of the eagerness to bring the software to the market. Of all the symptoms of poor software engineering, this is the one that management is most likely to condone.

Nevertheless, software engineering is becoming a true engineering discipline through the development of more rigorous approaches, the evangelization of standards, the nurturing of an accepted body of knowledge for the profession, and the proper education of software engineers. In many nations, the graduates of many accredited programs of software engineering are eligible to become registered professional engineers. In the United States a path to licensure for Professional Software Engineers existed from 2013 until 2020 but was abandoned due to lack of interest. The story of the rise and demise of software engineering licensure in the U.S. is interesting, informative, and also a cautionary tale involving the politics of technology and need for further professionalism among those developing mission-critical software (Laplante, 2020).

1.2.3 What Is the Difference between Software Engineering and Systems Engineering?

There is a great deal of similarity in the activities conducted in software and systems engineering. Table 1.1, adapted from an excellent introduction to software systems engineering by Richard Thayer, provides a summary of these activities. Take care when interpreting Table 1.1, as it has the tendency to suggest that the software engineering process is strictly a linear sequential (Waterfall) one. Various models of

TABLE 1.1
System Engineering Functions Correlated to Software System Engineering

System Engineering Function	Software Engineering Function	Software Engineering Description
Problem definition	Requirements analysis	Determine needs and constraints by analyzing system requirements allocated to software
Solution analysis	Software design	Determine ways to satisfy requirements and constraints, analyze possible solutions, and select the optimum one
Process planning	Process planning	Determine product development tasks, precedence, and potential risks to the project
Process control	Process control	Determine methods for controlling project and process, measure progress, and take corrective action where necessary
Product evaluation	Verification, validation, and testing	Evaluate final product and documentation

Source: Adapted from Thayer, 2002.

software development will be discussed in the next chapter. Also, note that there is no mention of "coding" in Table 1.1. This is not an inadvertent omission. The writing of code can be the least engineering-like activity that a software engineer can undertake.

1.2.4 WHAT IS THE HISTORY OF SOFTWARE ENGINEERING?

Although early work in software development and software engineering began in the late 1950s, many believe that software engineering first became a profession as a result of a NATO-sponsored conference on software engineering in 1968. It is certainly true that this conference fueled a great deal of research in software engineering (Marciniak, 1994).

The 1970s were a time when software engineering began its rise as new ideas, languages, and hardware were introduced. The 1980s continued to show great changes as the Software Crisis began to wind down. New languages and tools helped begin the journey toward better engineering and the move toward object-oriented programming begins. Major university programs in software engineering started to emerge in the late 1980s.

The 1990s decade was a boom for programming languages, with some of the most popular ones used today being introduced. In 1993, the ACM/IEEE Steering Committee for the Establishment of Software Engineering as a Profession recommended the following:

- Adopt standard definitions
- Define a required body of knowledge and recommended practices

- Define ethical standards
- Define educational curricula3 as a means toward greater uniformity and professionalism in software engineering.

These recommendations, and an extensive collaborative project over many years, led to the creation of the first Software Engineering Body of Knowledge (SWEBOK), sponsored by the Institute for Electrical and Electronics Engineers (IEEE) Computer Society (Laplante, 2020).

A number of other big changes to the software engineering industry also occurred in this decade: object-oriented programming began to grow in popularity, the Internet made its debut, and a new approach to development was introduced.

The greater focus in the 2000s was on methodology as developers looked to make the process more responsive to customer needs, more profitable, and easier to create.

While providing continuous improvements to languages and methods, the focus in the 2010s shifted to addressing the need for software engineers with a new style of learning made to enhance traditional software engineering education.

1.2.5 WHAT IS THE ROLE OF THE SOFTWARE ENGINEER?

The production of software is a problem-solving activity that is accomplished by modeling. As a problem-solving, modeling discipline, software engineering is a human activity that is biased by previous experience and is subject to human error. Therefore, the software engineer should recognize and try to eliminate these errors.

Software engineers should also strive to develop code that is built to be tested, that is designed for reuse, and that is ready for inevitable change. Anticipation of problems can only come from experience and from drawing upon a body of software practice experience that is more than 60 years old.

1.2.6 HOW DO SOFTWARE ENGINEERS SPEND THEIR TIME ON THE JOB?

Software engineers probably spend less than 10% of their time writing code. The other 90% of their time is involved with other activities, including:

1. Eliciting requirements
2. Analyzing requirements
3. Writing software requirements documents
4. Building and analyzing prototypes
5. Developing software architecture and designs
6. Writing software design documents
7. Researching software engineering techniques or obtaining information about the application domain
8. Developing test strategies and test cases
9. Testing the software and recording the results
10. Isolating problems and solving them
11. Learning to use or installing and configuring new software and hardware tools

12. Writing documentation such as users manuals
13. Attending meetings with colleagues, customers, and supervisors
14. Archiving software or readying it for distribution.

This is only a partial list of software engineering activities. These activities are not necessarily sequential and are not all-encompassing. Finally, most of these activities can recur throughout the software life cycle and in each new minor or major software version. Many software engineers specialize in a small subset of these activities, for example, software testing.

1.2.7 WHAT IS THE DIFFERENCE BETWEEN A SOFTWARE ENGINEER AND A SYSTEMS DEVELOPER?

The terms software engineer and software developer are frequently interchangeable. While the two roles may overlap often, a software engineer is a professional who applies all aspects of how to build software for a project as outlined above. That can include the design, maintenance, testing, and even evaluation for continuous improvement of the software. A software developer, on the other hand, is the professional who builds the software and makes sure it does what it's supposed to do.

1.2.8 WHAT KIND OF EDUCATION DO SOFTWARE ENGINEERS NEED?

Ideally, software engineers will have an undergraduate degree in software engineering, computer science, or electrical engineering with a strong emphasis on software systems development. While it is true that computer science and software engineering programs are not the same, many computer science curricula incorporate significant courses on important aspects of software engineering. Today, there are many undergraduate programs in software engineering.

An alternative path to the proper education in software engineering would be an undergraduate degree in a technical discipline and a Master's degree in software engineering (such as the degree that we are involved with at Penn State). Yet another path would be any undergraduate degree, significant experiential learning in software engineering on the job, and an appropriate Master's degree.

There are also numerous certifications in various technologies, and subdisciplines of software engineering such as security. These certifications may be offered by corporations, universities and professional organizations. Obtaining these certifications generally involves study, practice, testing, and/or skills demonstrations that form an important part of the continuing education required of every software professional.

Another aspect of education involves the proper background in the domain area in which the software engineer is practicing. So, a software engineer building medical software systems would do well to have significant formal education in the health sciences, biology, medicine, and the like. Again, continuing education and certification can be also helpful in obtaining and maintaining domain knowledge across disciplines.

1.2.9   WHAT KIND OF EDUCATION DO SOFTWARE ENGINEERS TYPICALLY HAVE?

Here is where the problem occurs. In our experience, many practicing software engineers have little or no formal education in software engineering. While software engineering is a discipline in which practice and experience are important, it is rare for someone who has not studied software engineering to have the skills and knowledge needed to efficiently and regularly build industrial strength, reliable software systems.

1.2.10   I KNOW SOMEONE WHO IS AN "XYZ CERTIFIED ENGINEER"; ARE THEY A SOFTWARE ENGINEER?

They may not be a software engineer. There is an important distinction between certification and licensing. Certification is a voluntary process administered by a nongovernment entity. Licensing is a mandatory process controlled by a state licensing board. Some companies tend to avoid the use of the word "engineer" in the designation because the courts generally rule in favor of restricting the use of the term.

Of course, if that person has other qualifications such as an undergraduate or graduate degree in software engineering, some experience and also holds the certification, then of course, they are a software engineer. As previously mentioned, certification can be an important part of the continuing education of every software engineering professional.

1.2.11   WHY ARE THERE SO MANY SOFTWARE ENGINEERS WITHOUT THE PROPER EDUCATION?

Software engineering is quite new, compared to other professions. The field barely existed 50 years ago, and it was only about 40 years ago that computers became accessible to people who didn't already work for large corporations. Shortages of trained software engineers in the 1980s and 1990s led to aggressive hiring of many without formal training in software engineering. This situation is commonly found in companies building engineering products where the "software engineers" were probably trained in some other technical discipline (e.g., electrical engineering, physics, or mechanical engineering) but not in software engineering. In other cases, there is a tendency to move technicians into programming jobs as instrument interfaces move from hardware- to software-based.

This situation has been changing, of course. Lots of software engineers with degrees are available, so companies are less willing to risk hiring those without unless they're able to prove themselves elsewhere first.

1.2.12   CAN SOFTWARE ENGINEERING PROGRAMS BE ACCREDITED?

Yes, a body known as CSAB can accredit undergraduate software engineering programs. The acronym CSAB formerly stood for Computing Sciences Accrediting Board but is now used without elaboration. CSAB is a participating body of ABET (formerly known as the Accreditation Board for Engineering and Technology but also known only by its acronym now). ABET accredits other kinds of undergraduate

TABLE 1.2
Most In-Demand Certificates for Software Engineers

Certificate	Provider
Software Engineering Master Certification (SEMC)	IEEE
Amazon Web Services	Amazon
Oracle Certified Master	Oracle
Salesforce Administrator	Salesforce
Microsoft Certified Azure Solutions Architect	Microsoft
Certified Scrum Master	Scrum Alliance
Project Management Professional	Project Management Institute (PMI)
Certified Software Engineer	The Institute of Certification of Computing Professionals (ICCP)
CIW Web Development Professional	CIW
ISTQB Certified Tester Foundation Level (CTFL)	ASTQB

engineering programs. Within ABET, the Computing Accreditation Commission accredits programs in computer science and information systems, while the Engineering Accreditation Commission accredits programs in software engineering and computer engineering.

The relevant member societies of ABET for software engineering are the Association for Computing Machinery (ACM), the Institute of Electrical and Electronics Engineers (Computer Society), and the Association for Information Systems.

1.2.13 What Are Some of the Popular Certificates in Demand for Software Engineers?

Table 1.2 lists some of the most in-demand certifications at the time of writing this book.

Tell us about yourself: Your profession, current role, education, and experience: https://pennstate.qualtrics.com/jfe/form/SV_4YjbJLO7dbsHPLM

1.3 STANDARDS AND CERTIFICATIONS

1.3.1 Are There Standards for Software Engineering Practices, Documentation, and So Forth?

There are many, and we list them in Table 1.3 for your reference. The title of the standard is self-explanatory. If the titles are not recognizable now, they will be after you have read this book.

TABLE 1.3
Common Standards in Use in Software Engineering

Standard Number	Standard Name
IEEE 610	Standard Glossary of Software Engineering Terminology
IEEE 828	Configuration Management in Systems and Software Engineering
IEEE 829	Software Test Documentation
IEEE 1028	Software Reviews and Audits
IEEE 1016	Information Technology—Systems Design—Software Design Descriptions
IEEE 1074	Developing a Software Project Life Cycle Process
IEEE 1471	Software architecture/system architecture
IEEE 12207	Information Technology—Software life-cycle processes

Of course, there are many other standards issued by various organizations around the world covering various aspects of software engineering and computing sciences. The above selection is provided both for reference purposes and to illustrate the depth and breadth of the software engineering standards that exist.

1.3.2 What Is the Software Engineering Body of Knowledge?

Since 1993, the IEEE Computer Society and the ACM have been actively promoting software engineering as a profession, notably through their involvement in accreditation activities described before and in the development of a Software Engineering Body of Knowledge (abbreviated as SWEBOK but often pronounced as "sweebock"). In 2002, the IEEE Computer Society introduced the Certified Software Development Professional (CSDP) designation, with testing built on top of SWEBOK. The Software Engineering Body of Knowledge is an international standard ISO/IEC TR 19759:2005 that describes the "sum of knowledge within the profession of software engineering." The CSDP and SWEBOK continued to be refreshed over the years, until late 2014, when the former was replaced by the Professional Software Engineering Master (PSEM) and Professional Software Engineering Process Master (PSEPM) certifications (CSDP, 2022).

In late 2013, SWEBOK Version 3 was approved for publication and released. In 2016, the IEEE Computer Society kicked off the SWEBOK evolution effort to develop future iterations of the body of knowledge. The objectives of the Guide to the SWEBOK project are to:

> "characterize the contents of the Software Engineering Body of Knowledge; provide a topical access to the Software Engineering Body of Knowledge; promote a consistent view of software engineering worldwide; clarify the place of, and set the boundary of, software engineering with respect to other disciplines such as computer science, project management, computer engineering and mathematics; provide a foundation for curriculum development and individual certification and licensing material."

The published version of SWEBOK Version 3 has the following 15 knowledge areas within the field of software engineering:

- Software requirements
- Software design
- Software construction
- Software testing
- Software maintenance
- Software configuration management
- Software engineering management
- Software engineering process
- Software engineering models and methods
- Software quality
- Software engineering professional practice
- Software engineering economics
- Computing foundations
- Mathematical foundations
- Engineering foundations.

Software engineers must also be knowledgeable in specifics of their particular application domain. For example, avionics software engineers need to have a great deal of knowledge of aerodynamics; software engineers for financial systems need to have knowledge in the banking domain, and so forth.

1.3.3 Are There Any "Fundamental Theorems" of Software Engineering?

Software engineering has often been criticized for its lack of a rigorous, formalized approach. And in those cases where formalization is attempted, it is often perceived as artificial and hence ignored (or accused of being computer science, not software engineering). Even a few of our most respected colleagues seem to hold this view. But there are some results in computer science, mathematics, and other disciplines that, while rigorous, can be shown to be applicable in a number of practical software engineering settings.

Rigor in software engineering requires the use of mathematical techniques. Formality is a higher form of rigor in which precise engineering approaches are used. In the case of the many kinds of systems, such as real-time, formality further requires that there be an underlying algorithmic approach to the specification, design, coding, and documentation of the software. In every course we teach, we try to be rigorous. For example, in many courses we introduce finite state machines because students can readily see that they are practical yet formal.

It has been stated over and over again (without convincing proof) that high-tech jobs are leaving the U.S. partly because Americans are inadequately trained in mathematics as compared to other nationalities. Yet most people decry the need for mathematics in software engineering.

The most laudable efforts to justify the need for mathematics education in software engineering and computer science offer pedagogical arguments centering on the need to demonstrate higher reasoning and logical skills. While these arguments are valid, most students and critics will not be satisfied by them. Software engineering students (and computer science students) want to know why they must take calculus and discrete mathematics in their undergraduate programs because they often do not see uses for it. Denning (2003) makes a plea for "great principles of computing;" that is, the design principles (simplicity, performance, reliability, evolvability, and security) and the mechanics (computation, communication, coordination, automation, and recollection). But perhaps there are no such grand theories, but rather many simple rules. Here is a list of some of our favorites:

Bayes' Theorem
Böhm-Jacopini Rule
Cantor's Diagonal Argument
Chebyshev's Inequality
Little's Law
McCabe's Cyclomatic Complexity Theorem
von Neumann's Min-Max Theorem.

Bayes' Theorem provides the underpinning for a large body of artificial intelligence using Bayesian Estimation.

Böhm-Jacopini's Rule shows that all programs can be constructed using only a `goto` statement. This theory has important implications in computability and compiler theory, among other places.

Cantor's Diagonal Argument was used by mathematician Georg Cantor to show that that real numbers are uncountably infinite. But Cantor's Argument can also be used to show that the Halting Problem is undecidable; that is, there is no way to a priori prove that a computer program will stop under general conditions.

Chebyshev's Inequality for a random variable x with mean σ and standard deviation, is stated as

$$1 - \frac{1}{k^2} \leq P_{k^2}\left(\left|x - \mu\right| \geq k\sigma\right) \qquad (1.1)$$

That is, the probability that the random variable will differ from its mean by k standard deviations is $1 - \frac{1}{k^2}$. Chebyshev's Inequality can be used to make all kinds of statements about confidence intervals. So, for example, the probability that random variable x falls within two standard deviations of the mean is 75% and the probability that it falls within six deviations of the mean (Six Sigma) is about 99.99%. This result has important implications in software quality engineering.

Little's Law is widely used in queuing theory as a measure of the average waiting time in a queue. Little's Law also has important implications in computer performance analysis.

McCabe's Cyclomatic Complexity Theorem demonstrates the maximum number of linearly independent code paths in a program and is quite useful in testing theory and in the analysis of code evolution. These features are discussed later.

Finally, von Neumann's Min-Max Theorem is used widely in economics and optimization theory. Min-Max approaches can also be used in all kinds of software engineering optimization problems from model optimization to performance improvement.

Although there are other many mathematical concepts familiar to all engineers that could be introduced in our software engineering classes, the aforementioned ones can be easily shown to be connected to one or more very practical notions in software engineering. Still, it is true that the discipline of software engineering is lacking grand theory, such as Maxwell's Equations or the various Laws of Thermodynamics, or even something as simple as the Ideal Gas Law in chemistry.

1.4   MISCONCEPTIONS ABOUT SOFTWARE ENGINEERING

1.4.1   WHY IS SOFTWARE SO BUGGY AND UNRELIABLE?

It is unclear if software is more unreliable than any other complex engineering endeavor. While there are sensational examples of failed software, there are just as many examples of failed engineered structures, such as bridges collapsing, space shuttles exploding, nuclear reactors melting down, and so on.

To us, it seems that software gets a bad rap. Oftentimes when a project fails, software engineering is blamed, not the incompetence of the managers, inadequacy of the people on the project, or the lack of a clear goal.

In any case, you have to prove that software is more unreliable than any other kind of engineering system, and we have seen no compelling evidence to support that contention.

1.4.2   I WRITE SOFTWARE AS PART OF MY JOB; DOES THAT MAKE ME A SOFTWARE ENGINEER?

No! Anyone can call themselves a software engineer if they write code, but they are not necessarily practicing software engineering. To be a software engineer, you need more than a passing familiarity with most of the concepts in this book.

1.4.3   BUT ISN'T SOFTWARE SYSTEM DEVELOPMENT PRIMARILY CONCERNED WITH PROGRAMMING?

As mentioned before, 10% or less of the software engineer's time is spent writing code. Someone who spends the majority of their time generating code is more aptly called a "programmer." Just as wiring a circuit designed by an electrical engineer is not engineering, writing code designed by a software engineer is not an engineering activity.

1.4.4   CAN'T SOFTWARE TOOLS AND DEVELOPMENT METHODS SOLVE MOST OR ALL OF THE PROBLEMS PERTAINING TO SOFTWARE ENGINEERING?

This is a dangerous misconception. Tools, software or otherwise, are only as good as the wielder. Bad habits and flawed reasoning can just as easily be amplified by tools as corrected by them. While software engineering tools are essential and provide significant advantages, relying on them to remedy process or engineering deficiencies is naïve.

1.4.5 Isn't Software Engineering Productivity a Function of System Complexity?

While it is certainly the case that system complexity can degrade productivity, there are many other factors that affect productivity. Requirements stability, engineering skill, quality of management, and availability of resources are just a few of the factors that affect productivity.

1.4.6 Once the Software Is Delivered, Isn't the Job Finished?

No. At the very least, some form of documentation of the end product, as well as the process used, needs to be written. More likely, the software product will now enter a maintenance mode after delivery in which it will experience many recurring life cycles as errors are detected and corrected and features are added.

1.4.7 Aren't Errors an Unavoidable Side Effect of Software Development?

While it is unreasonable to expect that all errors can be avoided (as in every discipline involving humans), good software engineering techniques can minimize the number of errors delivered to a customer. The attitude that errors are inevitable can be used to excuse sloppiness or complacency, whereas an approach to software engineering that is intended to detect every possible error, no matter how unrealistic this goal may be, will lead to a culture that encourages engineering rigor and high-quality software.

FURTHER READING

Abran, A., Moore, J.W., Bourque, P., Dupuis, R., & Tripp, L. (2004). Software engineering body of knowledge. *IEEE Computer Society, Angela Burgess*, 25.

Bourque, P., & Fairley, R.E. (2014). *IEEE Computer Society. 2014. Guide to the Software Engineering Body of Knowledge (SWEBOK (R)): Version 3.0.* IEEE Computer Society Press, Washington DC.

Certified Software Development Professional Program (CDSP), *IEEE Computer Society.* www.computer.org/education/certifications, accessed March 2022.

Denning, P.J. (2003). Great principles of computing. *Communications of the ACM, 46*(11), 15–20.

Laplante, P.A. (Editor-in-Chief) (2001). *Comprehensive Dictionary of Computer Science, Engineering and Technology.* CRC Press, Boca Raton, FL.

Laplante, P.A. (2005). Professional licensing and the social transformation of software engineers. *Technol. Soc. Mag., IEEE, 24*(2), 40–45.

Laplante, P.A. (2010). *Encyclopedia of Software Engineering. Auerbach Publications.* Software Engineering Body of Knowledge (SWEBOK). www.computer.org/education/bodies-of-knowledge/software-engineering, accessed March 2022.

Laplante, P.A. (2020). A brief history of software professionalism and the way forward. *Computer.* vol *53*, no. 9: pp. 97–100.

Marciniak, J. (Ed) (1994). *Encyclopedia of Software Engineering, Vol. 2.* John Wiley & Sons, pp. 528–532.

Thayer, R.H. (2002). Software system engineering: a tutorial. *Computer, 35*(4), 68–73.

Tripp, L.L. (2002). Benefits of certification. *Computer, 35*(6), 31–33.

2 Software Properties, Processes, and Standards

OUTLINE

- Characteristics of software
- Software processes and methodologies
- Software standards

2.1 INTRODUCTION

To paraphrase Lord Kelvin, if you can't measure that which you are talking about, then you really don't know anything about it. In fact, one of the major problems with portraying software engineering as a true engineering discipline is the difficulty with which we have in characterizing various attributes, characteristics, or qualities of software in a measurable way. We begin this chapter, then, with the quantification of various attributes of the software.

Every software process is an abstraction, but the activities of the process need to be mapped to a life cycle model. There is a variety of software life cycle models, which are discussed in this chapter. We significantly focus coverage on the activities of the agile approaches which have been dominating the landscape of software development lifecycles for many years now.

We conclude the chapter by discussing some of the previously mentioned software standards that pertain to software qualities, life cycles, and processes.

2.2 CHARACTERISTICS OF SOFTWARE

2.2.1 HOW DO SOFTWARE ENGINEERS CHARACTERIZE SOFTWARE?

Software systems are characterized both by their functional behavior (what the software does) and by a number of qualities (how the software behaves concerning some observable attributes like reliability, reusability, maintainability, etc.). In the software marketplace, in which functionally equivalent products compete for the same customer, quality attributes become more important in distinguishing between the competing products.

2.2.2 WHAT IS SOFTWARE QUALITY?

Quality is the totality of characteristics of an entity that bear on its ability to satisfy stated and implied needs. Software quality is an essential and distinguishing attribute

DOI: 10.1201/9781003218647-3

of the final software product. Evaluation of software products in order to satisfy software quality needs is one of the processes in the software development life cycle.

Software product quality can be evaluated by measuring internal quality attributes, or by measuring external quality attributes. Internal qualities are those that may not necessarily be visible to the user but help the developers to achieve improvement in external qualities. For example, good requirements and design documentation might not be seen by the typical user, but these are necessary to achieve improvement in most of the external qualities. On the other hand, the external qualities, such as security, usability, and reliability, are visible to the user. A specific distinction between whether a particular quality is external or internal is not often made because they are so closely tied. Moreover, the distinction is largely a function of the software itself and the kind of user involved.

2.2.3 WHAT ARE THE CONSEQUENCES OF IGNORING QUALITY ATTRIBUTES WHILE BUILDING SOFTWARE?

Once the software has been deployed, it is typically straightforward to observe whether a certain functionality has been met, as the areas of success or failure in their context can be rigidly defined. However, the same is not true for quality attributes as these can refer to concepts that can be interdependent and difficult to measure. The problem of lacking integration of any early quality attributes within the development process is likely to cause an increase in the effort and maintenance overhead, or even a catastrophic project failure. The following list provides valid examples:

- Target Data Breach (O'Neil, 2015): Between November 27 and December 18, 2013, the Target Corporation's network was breached, which became the second-largest credit and debit card breach after the TJX breach in 2007. In the Target incident, 40 million credit and debit card numbers were stolen in only two and a half weeks. Target's systems were compromised by a stolen password. Upon analyzing the security breach, it was found that an HVAC company had access to their networks to monitor the environmental conditions of the servers. The perpetrators were then able to legitimately gain access to Target's systems through the available access that was allowed for the HVAC company. Target eventually installed malware on the Point of Sale (POS) systems.
- Siemens: Possible Hearing Damage in Some Cell Phones (Siemens, 2004): In 2004, Siemens issued a safety warning that some of its cell phones may have a software problem that could cause them to emit a loud noise, possibly causing a hearing loss for the phone user. The malfunction happened only if, while the phone was in use, the battery ran down to the point that the phone automatically disconnected the call and began to shut down.
- Mars Climate Orbiter (Breitman, Leite, & Finkelstein, 1999): This was one of two NASA spacecraft in the Mars Surveyor '98 program. The mission failed because of a software interoperability issue. The craft drifted off course during its voyage and entered a much lower orbit than planned and was destroyed by atmospheric friction. The metric/imperial mix-up, which destroyed the craft,

was caused by a software error on Earth. The thrusters on the spacecraft, which were intended to control its rate of rotation were controlled by a terrestrial computer that underestimated the effect of the thrusters by a factor of 4.45. This is the ratio between a pound-force – the standard unit of force in the imperial system – and a Newton, the standard unit in the metric system. The software on Earth was working in pounds-force, while the spacecraft expected figures in Newtons.

- The National Library of Medicine MEDLARS II system (Boehm & In, 1996): The project was initially developed with many layers of abstraction to support a wide range of future publication systems. The initial focus of the system was toward improving "portability" and "evolvability" qualities. The system was scrapped after two expensive hardware upgrades due to performance problems.

- The New Jersey Department of Motor Vehicles' licensing system (Babcock, 1985): This system was written in a fourth-generation programming language, ideal to save development time. When implemented, the system was so slow that, at one point, more than a million New Jersey vehicles roamed the streets with unprocessed license renewals. The project aimed at satisfying affordability and timeliness objectives but failed due to performance scalability problems.

2.2.4 WHAT ARE THE DIFFERENT QUALITIES THAT CAN CHARACTERIZE SOFTWARE?

Many approaches and taxonomies exist that aim to catalog and classify software quality into a structured set of characteristics that are further decomposed into sub-characteristics, for example, Boehm (1976), Chung, Nixon, Yu, & Mylopoulos (2000), ISO 25000, and Kassab (2009).

In the following discussion we review some of the most widely used quality attributes; namely:

- Security
- Performance
- Usability
- Portability
- Interoperability
- Testability
- Modifiability
- Reliability
- Correctness
- Traceability.

2.2.5 WHAT IS SOFTWARE "SECURITY"?

To be able to argue about security, one has to understand what it means. Let's examine security in the light of three contexts:

- Think about the Fort Knox U.S. Army base in Kentucky. In addition to housing various U.S. Army functional units, it is also the home to a gold bullion depository housing 5,000+ tons of gold. So, what is the business asset that needs security in this context? And what does "security" mean in this context?
- Now consider the U.S. Central Intelligence Agency (CIA), which is a federal intelligence organization whose primary purpose is to collect information about foreign governments, corporations, and individuals. The CIA uses this information to inform and possibly influence public policymakers. In this case, what is the business asset that needs to be secured? And what does "security" mean in this case?
- Now, think of an electrical distribution grid system. Again, what is the asset that needs to be secured? And what does "security" mean here?

To define security in a specific context, we need to understand the asset that requires security or protection. We then want to ensure that the system protects that asset and we do this by imagining how the asset is at risk. Then we then articulate the desired response to that risk.

So, security is defined as a measure of the system's ability to protect data and information from unauthorized access while still providing access to people and systems that are authorized.

An action taken against a computer system with the intention of doing harm is called an attack and can take a number of forms. The attack may be an unauthorized attempt to access data or services or to modify data, or it may be intended to deny services to legitimate users. The related concerns that refine security are typically classified as security concerns. These concerns are typically:

- Confidentiality: The property that reflects the extent to which data and services are only available to those that are authorized to access them.
- Integrity: It reflects the extent to which data or services can be delivered as intended. This property can also refer to data or services.
- Non-Repudiation: It refers to the ability to guarantee a sender of a message from the system cannot later repudiate or deny having sent the message. It can also refer to the guarantee that the recipient cannot later deny having received the message.
- Availability: This is the property that reflects the extent to which the system will be available for legitimate use. A denial-of-service attack is meant to disrupt the availability of a system, and it is a security concern. Availability builds on reliability by adding the notion of recovery (repair). Fundamentally, availability is about minimizing service outage time by mitigating faults.

Not all of these four security concerns are of the same importance to every system. For example, the privacy concern is not necessarily of the same importance for a security system in a public museum as for an accounts management system in a financial institution.

People often have trouble articulating security concerns and it is common to articulate the solution (e.g., "System must support login") instead of the concern itself.

Here "Support login" is a functional requirement that may emerge as a solution to satisfy a confidentiality concern.

2.2.6 WHAT IS SOFTWARE "PERFORMANCE"?

Performance is about time and the system's ability to meet timing requirements. When events occur—interrupts, messages, requests from users or other systems, or clock events marking the passage of time—the system, or some element of the system, must respond to them in a timely way. Characterizing the events that can occur (and when they can occur) and the system or element's time-based response to those events is the essence of a discussion of a system's performance. For example, in a Home Automation System, in case of a fire, we expect the system to generate a notifying alarm within 60 seconds.

2.2.7 WHAT IS SOFTWARE "USABILITY"?

Usability is a measure of how easy the software is for humans to use. Software usability is synonymous with ease-of-use, or user-friendliness. Properties that make an application user-friendly to novice users are often different from those desired by expert users or software designers. The use of prototyping can increase the usability of a software system because, for example, interfaces can be built and tested by the user.

Usability quality is an elusive one. Usually, informal feedback from users is used, as well as surveys, focus groups, and problem reports. "Designer as an apprentice", a requirements discovery and refinement technique that will be discussed in Chapter 3, can also be used to determine usability.

2.2.8 WHAT IS SOFTWARE "PORTABILITY"?

Software is portable if it can run easily in different environments. The term environment refers to the hardware on which the system resides, the operating system, or other software in which the system is expected to interact. The Java programming language, for example, was invented to provide a program execution environment that supported full portability across a wide range of embedded systems platforms and applications (see Chapter 6).

Portability is difficult to measure, other than through anecdotal observation. Person months required to perform the port is the standard measure of this property. Portability is achieved through a deliberate design strategy in which hardware-dependent code is confined to the fewest code units as possible. This strategy can be achieved using either object-oriented or procedural programming languages and through object-oriented or structured approaches. All of these will be discussed later.

2.2.9 WHAT IS SOFTWARE "INTEROPERABILITY"?

Interoperability is about the degree to which two or more software systems can usefully exchange meaningful information. Like all quality attributes, interoperability is not a yes-or-no proposition but has shades of meaning. Interoperability can be measured in terms of compliance with open system standards. These standards are typically specific to the application domain. For example, in the railway industry, the prevailing standard of interoperability is IEEE 1473 – 1999 (IEEE, 1999).

2.2.10 WHAT IS SOFTWARE "TESTABILITY"?

Testability refers to the ease with which software or system can be made to demonstrate its faults through (typically execution-based) testing. Specifically, testability refers to the probability, assuming that the software or system has at least one fault, that it will fail on its next test execution. If a fault is present in a system, then we want it to fail during testing as quickly as possible. For a system to be properly testable, it must be possible to control each component's inputs (and possibly manipulate its internal state) and then to observe its outputs (and possibly its internal state). One common technique for increasing testability is through the insertion of software code that is intended to monitor various qualities such as performance or correctness. Modular design (to be discussed in Chapter 5), rigorous software engineering practices, and the effective use of an appropriate programming language can also contribute to testability.

2.2.11 WHAT IS SOFTWARE "MODIFIABILITY"?

Modifiability is about change and the software engineer's interest in it is in the cost and risk of making changes. To plan for modifiability, a software engineer must consider three questions:

- What can change?
- What is the likelihood of the change?
- When is the change made and who makes it?

2.2.12 WHAT IS SOFTWARE "RELIABILITY"?

Software reliability can be defined informally in a number of ways. For example, can the user "depend on" the software? Other loose characterizations of a reliable software system include:

- The system "stands the test of time."
- There is an absence of known catastrophic errors (those that disable or destroy the system).
- The system recovers "gracefully" from errors.
- The software is robust.

For engineering-type systems, other informal views of reliability might include the following:

- Downtime is below a certain threshold.
- The accuracy of the system is within a certain tolerance.
- Real-time performance requirements are met consistently.

2.2.13 How Do You Measure Software Reliability?

Software reliability can be defined in terms of statistical behavior; that is, the probability that the software will operate as expected over a specified time interval. These characterizations generally take the following approach. Let S be a software system and let T be the time of system failure. Then the reliability of S at time t, denoted $r(t)$, is the probability that T is greater than t; that is,

$$r(t) = P(T > t) \tag{2.1}$$

This is the probability that a software system will operate without failure for a specified period.

Thus, a system with the reliability function $r(t) = 1$ will never fail. However, it is unrealistic to have such expectations. Instead, some reasonable goal should be set. For example, in a baggage inspection system at an airport, a reasonable standard of reliability might be that the failure probability be no more than 10^{-9} per hour. This represents a reliability function of $r(t) = (0.99999999)^t$ with t in hours. Note that as $t \to \infty$, $r(t) \to 0$.

2.2.14 What Is a Failure Function?

Another way to characterize software reliability is in terms of a real-valued failure function. One failure function uses an exponential distribution where the abscissa is time and the ordinate represents the expected failure intensity at that time (Equation 2.2).

$$f(t) = \lambda e^{-\lambda t} \quad t \geq 0 \tag{2.2}$$

Here the failure intensity is initially high, as would be expected in new software as faults are detected during testing. However, the number of failures would be expected to decrease with time, presumably as failures are uncovered and repaired (Figure 2.1). The factor λ is a system-dependent parameter.

2.2.15 What Is a "Bathtub Curve"?

The bathtub curve (see Figure 2.2) is often used to explain the failure function for physical components that wear out, electronics, and even biological systems. Obviously, we expect a large number of failures early in the life of a product (from manufacturing defects) and then a steady decline in failure incidents until later in the life of that product when it has worn out or, in the case of biological entities, died. But Brooks (1995) notes that the bathtub curve might also be useful in describing the number of errors found in a certain release of a software product.

2.2.16 But Software Doesn't Wear Out, So Why Would It Conform to the Bathtub Curve?

It is clear that a large number of errors will be found in a particular software product early, followed by a gradual decline as defects are discovered and corrected, just as

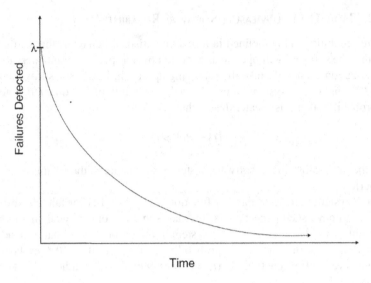

FIGURE 2.1   An exponential model of failure represented by the failure function $f(t) = \lambda e^{-\lambda t}$, $t \geq 0$. λ is a system-dependent parameter

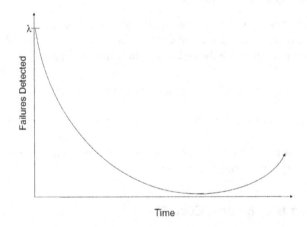

FIGURE 2.2   A software failure function represented by the bathtub curve

in the exponential model of failure. But we have to explain the increase in failure intensity later in time. There are at least three possible explanations. The first is that the errors are due to the effects of patching the software for bug fixes or new features.

The second reason for a late surge in failures is that the underlying hardware or operating system may have recently changed in a way that the software engineers did not anticipate, or the hardware may have failed due to age.

Finally, additional failures could appear because of the increased stress on the software by expert users. That is, as users master the software and begin to expose and strain advanced features, it is possible that certain poorly tested functionality of the software is beginning to be used.

2.2.17 CAN THE TRADITIONAL QUALITY MEASURES OF MTFF OR MTBF BE USED TO STIPULATE RELIABILITY IN THE SOFTWARE REQUIREMENTS SPECIFICATION?

Yes, mean time to first failure (MTFF) which measures the amount of time on average that a part can run before it breaks, and MTBF which measures the time between failures, both can be used. This approach to failure definition places great importance on the effective elicitation (discovery) and specification of functional requirements because the requirements define the software failure.

2.2.18 WHAT IS MEANT BY THE "CORRECTNESS" OF SOFTWARE?

Software correctness is closely related to reliability and the terms are often used interchangeably. The main difference is that a minor deviation from the requirements is strictly considered a failure and hence means the software is incorrect. However, a system may still be deemed reliable if only minor deviations from the requirements are experienced. Correctness can be measured in terms of the number of failures detected over time.

2.2.19 WHAT IS "TRACEABILITY" IN SOFTWARE SYSTEMS?

Traceability is concerned with the relationships between requirements, their sources, and the system design. Regardless of the process model, documentation and code traceability is paramount. A high level of traceability ensures that the software requirements flow down through the design and code and then can be traced back up at every stage of the process. This would ensure, for example, that a coding decision can be traced back to a design decision to satisfy a corresponding requirement.

Traceability is particularly important in embedded systems because often design and coding decisions are made to satisfy hardware constraints that may not be easily associated with a requirement. Failure to provide a traceable path from such decisions through the requirements can lead to difficulties in extending and maintaining the system.

Generally, traceability can be obtained by providing links between all documentation and the software code. In particular, there should be links:

- From requirements to stakeholders who proposed these requirements
- Between dependent requirements
- From the requirements to the design
- From the design to the relevant code segments
- From requirements to the test plan
- From the test plan to test cases.

2.2.20 ARE THERE OTHER SOFTWARE QUALITIES?

Of course, there are many software qualities that could be discussed, some mainstream, others more esoteric or application specific. While classic quality attributes (e.g., availability, modifiability, performance, security, etc.) are well discussed in books and literature, in practice, there are other quality attributes that are not so familiar as they are domain-specific or may emerge in the context of disruptive technology. Here are some examples of more recently characterized qualities:

- Caring: This quality could mean different things to different people and different systems. Consider, for example, a robotic surgery system. These systems are now used extensively for many types of procedures including heart, cancer, and prostate surgery. While current systems are robotic in the sense that the machine mimics the movements of a human surgeon, fully autonomous robot surgical systems are envisioned in the near future, replacing surgeons and nurses in the operating room.

   While we expect the human surgeon and nurses to care about the patient, as software engineers, what should we require of a fully autonomous robot surgeon? Caring likely encompasses elements of the qualities of trust, reliability, privacy, and more, but none of these, by themselves, capture the full essence of caring. Instead, caring is a "super-ility" resulting as some composite of safety, trust, reliability, privacy (and another new software quality, "empathy").

   Consider the constituent qualities of caring in the robotic surgery system. Likely, the surgeon wants the system to be safe and reliable as the primary concern. Both the safe and trustworthy operation of the system contributes to a sense of reliability in the system and are of concern to the systems engineers. The patient shares these concerns but also wants the system to preserve his privacy (e.g., by not exposing medical records or embarrassing images). If the actors in the operating room were humans, the patient would probably also expect a sense of empathy from the surgeons or nurses. Of course, robot surgeons look nothing like human surgeons, therefore there would need to be a means by which the robots could emote empathy via speech or facial expression generation on some display device. These diverse concerns, with respect to the qualities related to caring, will inform the specific system requirements discovery and representation process. Laplante, Kassab, Laplante, & Voas (2017) discuss the caring quality in the context of Internet of Things (IoT) in the healthcare domain.

- Humanization: There are questions on the moral role that IoT may play in human lives, particularly concerning personal control. Applications in the IoT involve more than computers interacting with other computers. Fundamentally, the success of the IoT will depend less on how far the technologies are connected and more on the humanization of the technologies that are connected. IoT technology may reduce people's autonomy, shift them toward habits, and then shift power to corporations focused on financial gain.

   For the education domain, for example, this effectively means that the controlling agents are the organizations that control the tools used by the academic professionals but not the academic professionals themselves. Dehumanization of humans in interacting with machines is a valid concern and it is discussed in many studies (e.g., Kassab. DeFranco, & Voas, 2018; Kassab, DeFranco, & Laplante, 2020). Many studies indicate that face-to-face interaction between students will not only benefit a child's social skills but also positively contribute to character building. The issue that may arise from increased IoT technologies in education is the partial loss of the social aspect of going to school. Hence, considering "Humanization" as a quality for any such systems worth the research.

- Iowability: In the book "Software Architecture in Practice" (Bass, Clements, & Kazman, 2021), the authors refer to a system architecture that was designed with the conscious goal of retaining key employees and attracting talented new hires to relocate to a quiet region of the American Midwest. That system's architects spoke of imbuing the system with "Iowability". This quality attribute was achieved for that system by bringing in state-of-the-art technology and giving their development teams wide creative latitude. While it is unlikely to find this quality attribute in any standard list of quality attributes, it is still important to that system in that specific context. Quality attribute requirements are derived from the context, not a particular textbook.

For brevity, we have confined the discourse to the most commonly discussed qualities of software. An in-depth discussion of these and other qualities to be considered can be found throughout the literature, for example, Kassab (2009).

Supplemental Materials from the authors on non-functional requirements: https://phil.laplante.io/requirements/NFR/supplements.php

2.3 SOFTWARE PROCESSES AND METHODOLOGIES

2.3.1 What Is a Software Process?

A software process is a model that describes an approach to the production and evolution of software. Software process models are frequently called "life cycle" models, and the terms are interchangeable.

2.3.2 Isn't Every Software Process Model Just an Abstraction?

As with any model, a process model is an abstraction. But in this case, the model depicts the process of translation—from system concept to requirements specification, to a design, then code, then finally via compilation and assembly, to the stored program form.

2.3.3 What Benefits Are There to Using a Software Process Model?

A good process model will help minimize the problems associated with each translation. A software process also provides for a common software development framework both within a project and across projects. The process allows for productivity improvements and it provides for a common culture, a common language, and common skills among organizational members. These benefits foster a high level of traceability and efficient communication throughout the project. In fact, it is very difficult to apply correct project management principles when an appropriate process model is not in place.

2.3.4   WHAT IS A SOFTWARE METHODOLOGY?

The methodology describes the "how." It identifies how to perform activities for each period, how to represent the activities and products,[1] and how to generate products.

2.3.5   AREN'T SOFTWARE PROCESS MODELS AND METHODOLOGIES THE SAME?

A software methodology is not the same as a software process. A software process is, in essence, the "what" of the software product life cycle. The process identifies and determines the order of phases within the life cycle. It establishes phase transition criteria and indicates what is to be done in each phase and when to stop. However, the terms process model and methodology are often used interchangeably (and, possibly, incorrectly). For example, there is both an agile software process model and many different agile methodologies, to be discussed shortly.

2.3.6   WHAT IS THE WATERFALL LIFE CYCLE MODEL?

The terms waterfall, conventional, or linear sequential are used to describe a sequential model of nonoverlapping and distinctive activities related to software development. Collectively, the periods in which these activities occur are often referred to as phases or stages. While simplistic and dating back at least 30 years, the waterfall model is still popular. A recent survey, for example, showed that 21% of companies still use a waterfall model (Kassab & Laplante, 2022).

2.3.7   HOW MANY PHASES SHOULD THE WATERFALL MODEL HAVE?

The number of phases differs between variants of the model. As an example of a waterfall model, consider a software development effort with activity periods that occur in the following sequence:

- Concept definition
- Requirements specification
- Design specification
- Code development
- Testing
- Maintenance.

The waterfall representation of this sequence is shown in Figure 2.3. Table 2.1 summarizes the activities in each period and the main artifacts of these activities.

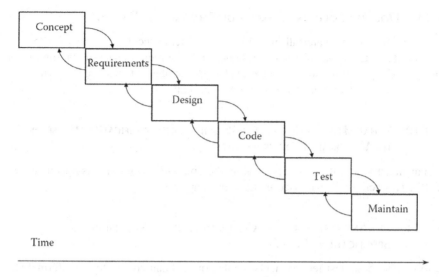

FIGURE 2.3 A waterfall life cycle model. The forward arcs represent time sequential activities. The reverse arcs represent backtracking

TABLE 2.1
Phases of a Waterfall Software Life Cycle with Associated Activities and Artifacts

Phase	Activity	Output
Concept	Define project goals	White paper
Requirements	Decide what the software must do	Software requirements specification
Design	Show how the software will meet the requirements	Software design description
Development	Build the system	Program code
Test	Demonstrate requirements satisfaction	Test reports
Maintenance	Maintain system	Change requests, reports

2.3.8 What Happens during the Software Conception Phase of the Waterfall Process Model?

The software conception activities include the determination of the software project's needs and overall goals. These activities are driven by management directives, customer input, technology changes, and marketing decisions.

At the onset of the phase, no formal requirements are written, generally, no decisions about hardware/software environments are made, and budgets and schedules cannot be set. In other words, only the features of the software product and possibly the feasibility of testing them are discussed. Usually, no documentation other than internal feasibility studies, white papers, or memos are generated.

2.3.9   Does the Software Conception Phase Really Happen?

Some variants of the waterfall model do not explicitly include a conception period because the activity was either incorporated into the requirements definition or not thought to be part of the software project at all. Nonetheless, the concept activity does occur in every software product, even if it is implicit.

2.3.10   What Happens during the Requirements Specification Phase of the Waterfall Process Model?

The main activity of this phase is creating the Software Requirements Specifications (SRS). This activity is discussed in detail in Chapter 3.

2.3.11   Do Any Test Activities Occur during the Requirements Specification Phase?

During this phase, test requirements are determined and committed to a formal test plan. The test plan is used as the blueprint for the creation of test cases used in the testing phase, which is discussed later in the text.

The requirements specification phase can and often does occur in parallel with product conception and, as mentioned before, they are often not treated as distinct. It can be argued that the two are separate, however, because the requirements generated during conceptualization are not binding, whereas those determined in the requirements specification phase are (or should be) binding. This rather subtle difference is important from a testing perspective because the SRS represents a binding contract and, hence, the criteria for product acceptance. Conversely, ideas generated during system conceptualization may change and, therefore, are not yet binding.

2.3.12   What Happens during the Software Design Phase of the Waterfall Process Model?

The main activity of software design is to develop a coherent, well-organized representation of the software system suitable to guide software development. In essence, the design maps the "what" from the SRS to the "how" of the software design description. Techniques for software design are discussed in Chapter 5.

2.3.13   Do Any Test Activities Occur during the Software Design Phase?

Certain test activities occur concurrently with the preparation of the software design description. These include the development of a set of test cases based on the test plan. Techniques for developing test cases are discussed in Chapter 7.

Often during the software design phase problems in the SRS are identified. These problems may include conflicts, redundancies, or requirements that cannot be met with current technology. In real-time systems the most typical problems are related to deadline satisfaction.

Usually, problems such as these require changes to the SRS or the granting of exemptions from the requirements in question. In any case, the problem resolution shows up as a specific directive in the software design description.

2.3.14 WHAT HAPPENS DURING THE SOFTWARE DEVELOPMENT PHASE OF THE WATERFALL PROCESS MODEL?

This phase involves the production of the software code based on the design using best practices. These best practices will be discussed in Chapter 6.

2.3.15 WHAT TEST ACTIVITIES OCCUR DURING THE SOFTWARE DEVELOPMENT PHASE?

During this phase the test team can build the test cases specified in the design phase in some automated form. This approach guarantees the efficacy of the tests and facilitates repeat testing.

2.3.16 WHEN DOES THE SOFTWARE DEVELOPMENT PHASE END?

The software development phase ends when all software units have been coded, unit tested, and integrated, and have passed the integration testing specified by the software designers.

2.3.17 WHAT HAPPENS DURING THE TESTING PHASE OF THE WATERFALL PROCESS MODEL?

Although ongoing testing is an implicit part of the waterfall model, the model also includes an explicit testing phase. These testing activities (often called acceptance testing to differentiate them from code unit testing) begin when the software development phase has ended. During the explicit testing phase, the software is confronted with a set of test cases (module and system level) developed in parallel with the software and documented in a software test requirements specification (STRS). Acceptance or rejection of the software is based on whether it meets the requirements defined in the SRS using tests and criteria set forth in the STRS.

2.3.18 WHEN DOES THE TESTING PHASE END?

The testing phase ends when either the criteria established in the STRS are satisfied, or failure to meet the criteria forces requirements modification, design alteration, or code repair. Regardless of the outcome, one or more test reports are prepared which summarize the conduct and results of the testing. More on testing, including test stoppage criteria, will be discussed in Chapter 7.

2.3.19   WHAT HAPPENS DURING THE SOFTWARE MAINTENANCE PHASE OF THE WATERFALL PROCESS MODEL?

The software maintenance phase activities generally consist of a series of reengineering processes to prolong the life of the system. Maintenance activities can be adaptive, which result from external changes to which the system must respond; corrective, which involves maintenance to correct errors; or perfective, which is all other maintenance including user enhancements, documentation changes, efficiency improvements, and so on. The maintenance activity ends only when the product is no longer supported.

In some cases, the maintenance phase is not incorporated into the life cycle model, but instead treated as a series of new software products, each with its own waterfall life cycle.

2.3.20   HOW ARE MAINTENANCE CORRECTIONS SUPPOSED TO BE HANDLED?

Maintenance corrections are usually handled by making a software change and then performing regression testing. Another approach is to collect a set of changes and then regression test against the last set of changes.

2.3.21   THE WATERFALL MODEL LOOKS ARTIFICIAL. IS THERE NO BACKTRACKING?

Yes, as shown in Figure 2.3, backtracking transitions do occur. For example, new features, lack of sufficient technology, or other factors force a reconsideration of the system's purpose. Redesign may result in a return to the requirements phase during design. Similarly, a transition from the programming phase back to the design phase might occur due to a feature that cannot be implemented or caused by an undesirable performance result. This in turn may necessitate redesign, new requirements, or elimination of the feature. Finally, a transition from the testing phase to the programming or design phases may occur due to an error detected during testing. Depending on the severity of the error, the solution may require reprogramming, redesign, modification of requirements, or reconsideration of the system goals.

2.3.22   WHAT IS THE V MODEL FOR SOFTWARE?

The V model is a variant of the waterfall model. It represents a tacit recognition that there are testing activities occurring throughout the waterfall software life cycle model and not just during the software testing period. These concurrent activities, depicted in Figure 2.4, are described alongside the activities occurring in the waterfall model.

For example, during requirements specification, the requirements are evaluated for testability and an STRS may be written. This document would describe the strategy necessary for testing the requirements. Similarly, during the design phase, a corresponding design of test cases is performed. While the software is coded and unit tested, the test cases are developed and automated. The test life cycle converges with the software development life cycle during acceptance testing.

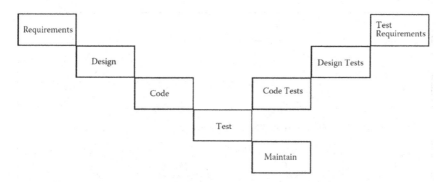

FIGURE 2.4 A V model for the software project life cycle. The concept phase is combined with the requirements phase in this instance

The point, of course, is that testing is a full life cycle activity and that it is important to constantly consider the testability of any software requirement and to design to allow for such testability.

2.3.23 WHAT IS THE SPIRAL MODEL FOR SOFTWARE?

The spiral model, suggested by Boehm (1988), recognizes that the waterfall model is not a realistic representation, nor is it necessarily a healthy one. Instead, the spiral model augments the waterfall model with a series of strategic prototyping and risk assessment activities throughout the life cycle. The spiral model is depicted in Figure 2.5.

Starting at the center of the figure, the product life cycle continues in a spiral path from the concept and requirements phases. Prototyping and risk analysis are used along the way to evaluate the feasibility of potential features. The added risk protection benefit from the extensive prototyping can be costly, but is well worth it, particularly in embedded systems.

More prototyping is used after a software development plan is written, and again after the design and tests have been developed. After that, the model behaves somewhat like a waterfall model.

2.3.24 WHAT ARE EVOLUTIONARY MODELS?

Evolutionary life cycle models promote software development by continuously defining requirements for new system increments based on experience from the previous version. Evolutionary models go by various names such as Evolutionary Prototyping, Rapid Delivery, Evolutionary Delivery Cycle, and Rapid Application Delivery.

In the evolutionary model, each iteration follows the waterfall model in that there are requirements, software design, and testing phases. After the final evolutionary step, the system enters the maintenance phase, although it can evolve again through the conventional flow, if necessary.

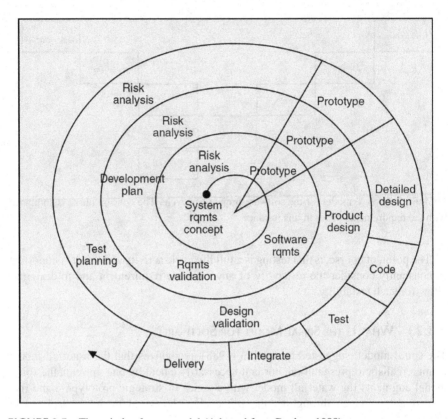

FIGURE 2.5 The spiral software model (Adapted from Boehm, 1988)

The evolutionary model can be used in conjunction with embedded systems, particularly in working with prototype or novel hardware that comes from simulators during development. Indeed, there may be significant benefits to this approach. First, early delivery of portions of the system can be generated, even though some of the requirements are not finalized. Then these early releases are used as tools for requirements elicitation, including timing requirements.

From the developers' point of view, those requirements that are clear at the beginning of the project drive the initial increment, but the requirements become clearer with each increment.

2.3.25 ARE THERE ANY DOWNSIDES TO USING EVOLUTIONARY MODELS?

Yes. For example, there may be difficulties in estimating costs and schedules when the scope and requirements are ill-defined. In addition, the overall project completion time may be greater than if the scope and requirements are established completely before design. Unfortunately, time apparently gained on the front end of a project because of early releases may be lost later because of the need for rework resulting from evolving requirements. Indeed, care must be taken to ensure that the evolving system architecture is both efficient and maintainable so that the completed system does not resemble a patchwork of afterthought add-ons. Finally, additional time must also be planned for integration and regression testing as increments are developed and added to the system. Some of the difficulties in using this approach in engineering systems can be mitigated, however, if the high-level requirements and overall architecture are established before entering an evolutionary cycle.

2.3.26 WHAT IS THE INCREMENTAL SOFTWARE MODEL?

The incremental model is characterized by a series of detailed system increments, each increment incorporating new or improved functionality to the system. These increments may be built serially or in parallel depending on the nature of the dependencies among releases and on availability of resources.

2.3.27 WHAT IS THE DIFFERENCE BETWEEN INCREMENTAL AND EVOLUTIONARY MODELS?

The difference is that the incremental model allows for parallel increments. In addition, the serial releases of the incremental model are planned whereas in the evolutionary model, each sequential release is a function of the experience from the previous iteration.

2.3.28 WHY USE THE INCREMENTAL MODEL?

There are several advantages to using the incremental model. These include ease of understanding each increment because of the decreased functionality, the use of successive increments in requirements elicitation, early development of initial functionality (which may aid in developing the real-time scheduling structure and for debugging prototype hardware), and successive building of operational functionality over time. The thinking is that software released in increments over time is more likely to satisfy changing user requirements than if the system were planned as a single overall release at the end of the same period. Finally, because the sub-projects are smaller, project management is more manageable for each increment.

2.3.29 ARE THERE ANY DOWNSIDES TO THE INCREMENTAL MODEL?

As with the evolutionary model, there may be increased system development costs as well as difficulties in developing temporal behavior and meeting timing constraints with a partially implemented system.

2.3.30 WHAT IS THE UNIFIED PROCESS MODEL?

The unified process model (UPM) uses an object-oriented approach by modeling a family of related software processes using the unified modeling language (UML) as a notation. Like UML, UPM is a metamodel for defining processes and their components.

The UPM consists of four phases, which undergo one or more iterations. In each iteration, some technical capability (software version or build) is produced and demonstrated against a set of criteria. The four phases in the UPM model are:

1. **Inception**: Establish software scope, use cases, candidate architecture, risk assessment.
2. **Elaboration**: Produce baseline vision, baseline architecture, select components.
3. **Construction**: Conduct component development, resource management and control.
4. **Transition**: Perform integration of components, deployment engineering, and acceptance testing.

Several commercial and open-source tools support the UPM and provide a basis for process authoring and customization.

2.3.31 WHERE IS THE UPM USED?

The UPM was developed to support the definition of software development processes specifically including those processes that involve or mandate the use of UML, such as the Rational Unified Process, and is closely associated with the development of systems using object-oriented techniques.

2.3.32 WHAT ARE AGILE METHODOLOGIES?

Agile methodologies comprise a family of nontraditional software development strategies that have captured the imagination of many who recognized the limitations of the traditional, process-laden (plan-driven) approaches. Agile methodologies are characterized by their lack of rigid process, though this fact does not mean that agile methodologies, when correctly employed, are not rigorous nor suitable for industrial applications—they are. What is characteristically missing from agile approaches, however, are "cook-book" solutions that focus on mandatory meetings, and complex documentation-prescribed development approaches.

Agile methodologies apply to software engineering. While there are elements of agile methodologies that can be applied to the engineering of systems (in particular, the human considerations), such methodologies are usually described as lightweight or lean when applied to non-software systems. This is so because agile methodologies depend on a series of rapid, non-throwaway prototypes, an approach that is not usually practical in hardware-based systems. In any case, the non-software engineer can still benefit from this chapter because agile methodologies are increasingly being employed and because the mindset of the agile software engineer includes some healthy perspectives.

In order to fully understand the nature of agile methodologies, we need to examine a document called the Agile Manifesto and the principles behind it. The Agile Manifesto was introduced by a number of leading proponents of agile methodologies to explain their philosophy (see Figure 2.6).

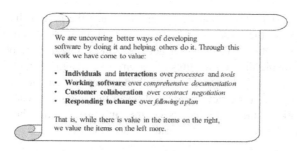

We are uncovering better ways of developing
software by doing it and helping others do it. Through this
work we have come to value:

- **Individuals** and **interactions** over *processes* and *tools*
- **Working software** over *comprehensive documentation*
- **Customer collaboration** over *contract negotiation*
- **Responding to change** over *following a plan*

That is, while there is value in the items on the right,
we value the items on the left more.

FIGURE 2.6 Manifesto for agile software development (From Beck, & Andres, 2004)

Signatories to the Agile Manifesto include many luminaries of modern software engineering practice such as Kent Beck, Mike Beedle, Alistair Cockburn, Ward Cunningham, Martin Fowler, Jim Highsmith, Ron Jeffries, Brian Marick, Robert Martin, Steve Mellor, and Ken Schwaber. Underlying the Agile Manifesto is a set of principles. Look at the principles below, noting the emphasis on those aspects that focus on requirements engineering, which we set in italics.

2.3.33 WHAT ARE THE PRINCIPLES BEHIND THE AGILE MANIFESTO?

- At regular intervals, the team reflects on how to become more effective, then tunes and adjusts its behavior accordingly.
- Our highest priority is to satisfy the customer through early and continuous delivery of valuable software.
- *Welcome changing requirements, even late in development.* Agile processes harness change for the customer's competitive advantage.
- Deliver working software frequently, from a couple of weeks to a couple of months, with a preference for the shorter timescale.
- *Business people and developers must work together daily throughout the project.*
- Build projects around motivated individuals. Give them the environment and support they need, and trust them to get the job done.
- *The most efficient and effective method of conveying information to and within a development team is face-to-face conversation.*
- Working software is the primary measure of progress.
- Agile processes promote sustainable development. The sponsors, developers, and users should be able to maintain a constant pace indefinitely.
- Continuous attention to technical excellence and good design enhances agility.
- Simplicity—the art of maximizing the amount of work not done—is essential. "Do the simplest thing that could possibly work."
- *The best architectures, requirements, and designs emerge from self-organizing teams.*

(Beck & Andres, 2004)

Notice how the principles acknowledge and embrace the notion that requirements change throughout the process. Also, the agile principles emphasize frequent, personal communication (this feature is beneficial in the engineering of non-software systems too). The highlighted features of requirements engineering in agile process models differ from the "traditional" waterfall and more modern models such as iterative, evolutionary, or spiral development. These other models favor a great deal of preliminary work on the requirements engineering process and the production of, often, voluminous requirements specifications documents.

2.3.34 What Are the Benefits of Agile Software Development?

Agile software development methods are a subset of iterative methods that focus on embracing change and emphasize collaboration and early product delivery, while maintaining quality. Working code is considered the true artifact of the development process. Models, plans, and documentation are important and have their value, but exist only to support the development of working software, in contrast with the other approaches already discussed. However, this does not mean that an agile development approach is a free-for-all. There are very clear practices and principles that agile methodologists must embrace.

- Agile methods are adaptive rather than predictive. This approach differs significantly from process models (e.g., Waterfall) that emphasize planning the software in great detail over a long period of time, and for which significant changes in the software requirements specification can be problematic. Agile methods are a response to the common problem of constantly changing requirements that can bog down the more "ceremonial" upfront design approaches, which focus heavily on documentation at the start.
- While in the waterfall model, testing is a separate phase that follows a build phase, in the agile models, testing is not sequential (in the sense it's executed only after coding phase) but continuous. What this also implies is that users in an agile environment can frequently use the newly developed components to validate the financial value of the software. After the users know the financial value of each iteration, they can make better decisions about the software's future.
- Agile methods are also people-oriented rather than process-oriented. This means they explicitly make a point of trying to make development "fun." Presumably, this is because writing software requirements specifications and software design descriptions is onerous and hence, to be minimized.

2.3.35 What Are Some Agile Methodologies?

Agile methodologies include Crystal, Extreme Programming, Scrum, Kanban, Lean, Dynamic Systems Development Method (DSDM), feature-driven development, adaptive programming, and there are others. We will look more closely at four of the most widely used agile methodologies: XP, Scrum, Kanban, and Lean.

2.3.36 WHAT IS EXTREME PROGRAMMING?

Extreme Programming (XP) is one of the most widely used agile methodologies. XP is traditionally targeted toward smaller development teams and requires relatively few detailed artifacts. XP takes an iterative approach to its development cycles. We can visualize the difference in process between a traditional waterfall model, iterative models, and XP (Figure 2.7). Whereas an evolutionary or iterative method may still have distinct requirements analysis, design, implementation, and testing phases similar to the waterfall method, XP treats these activities as being interrelated and continuous.

XP promotes a set of 12 core practices that help developers to respond to and embrace inevitable change. The practices can be grouped according to four practice areas:

- Planning
- Coding
- Designing
- Testing.

Some of the distinctive planning features of XP include holding daily stand-up meetings, making frequent small releases, and moving people through different project roles. Coding practices include having the customer constantly available, coding the unit test cases first, and employing pair-programming (a unique coding strategy where two developers work on the same code together).

Design practices include looking for the simplest solutions first, avoiding too much planning for future growth (speculative generality), and refactoring the code (improving its structure) continuously.

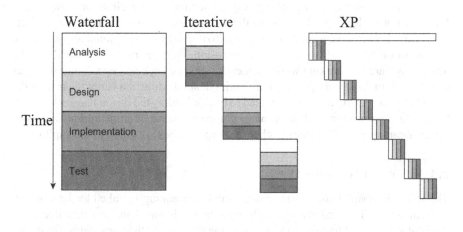

FIGURE 2.7 Comparison of waterfall, iterative, and XP development cycles (From Beck & Andres, 2004)

Testing practices include creating new test cases whenever a bug is found and having unit tests for all code, possibly using such frameworks as Xunit.

While removal of the territorial ownership of any code unit is another feature of XP, this framework may not work in the best way possible if all team members are not co-located.

2.3.37 What Is Scrum?

Scrum, which is named after a particularly contentious point in a rugby match, enables self-organizing teams by encouraging verbal communication across all team members and across all stakeholders. A fundamental principle of Scrum is that traditional, plan-driven software development methodologies such as waterfall and iterative development focus too much on process and not enough on software. Moreover, while plan-driven development focuses on non-software artifacts (e.g., documentation) and processes (e.g., formal reviews), Scrum emphasizes the importance of producing functioning software early and often. Scrum promotes self-organization by fostering high-quality communication between all stakeholders. In this case, it is implicit that the problem cannot be fully understood or defined (it may be a complex problem). And the focus in Scrum is on maximizing the team's ability to respond in an agile manner to emerging challenges.

Scrum features a "living" (constantly changing) backlog of prioritized work to be completed. Completion of a largely fixed set of backlog items occurs in a series of short (approximately 30 days) iterations or sprints. Each day, a brief (e.g., 15 minutes) meeting or Scrum is held in which progress is explained, upcoming work is described, and impediments are raised. A short planning session occurs at the start of each sprint to define the backlog items to be completed. A brief postmortem or heartbeat retrospective review occurs at the end of the sprint (Figure 2.8).

A Scrum Master removes obstacles or impediments to each sprint. The Scrum Master is not the leader of the team (as they are self-organizing) but acts as a productivity buffer between the team and any destabilizing influences. In some organizations, the role of the Scrum Master can cause confusion. For example, if two members of a Scrum team are not working well together, it might be expected by a senior manager that the Scrum Master fix the problem. Fixing team dysfunction is not the role of the Scrum Master. Personnel problems need to be resolved by the line managers to which the involved parties report. This scenario illustrates the need for institution-wide education about agile methodologies when such approaches are going to be employed.

2.3.38 What Is Kanban?

The term "Kanban" is of Japanese origin and its meaning is linked to the concept, "just in time". The underlying Kanban method originated in lean manufacturing (presented next) and focuses on visualizing the progress of the work items, from start to finish, usually through a Kanban board that can be physical or digital (Figure 2.9). Having the work progress visualized allows team members to see the state of every piece of work at any time.

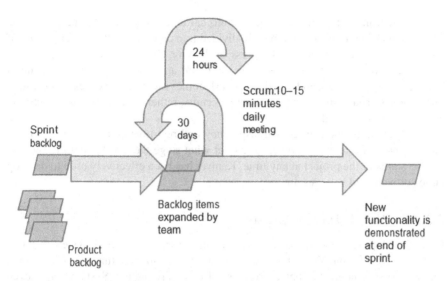

FIGURE 2.8   The Scrum development process from Boehm and Turner,2004 (Adapted from Schwaber & Beedle, 2001)

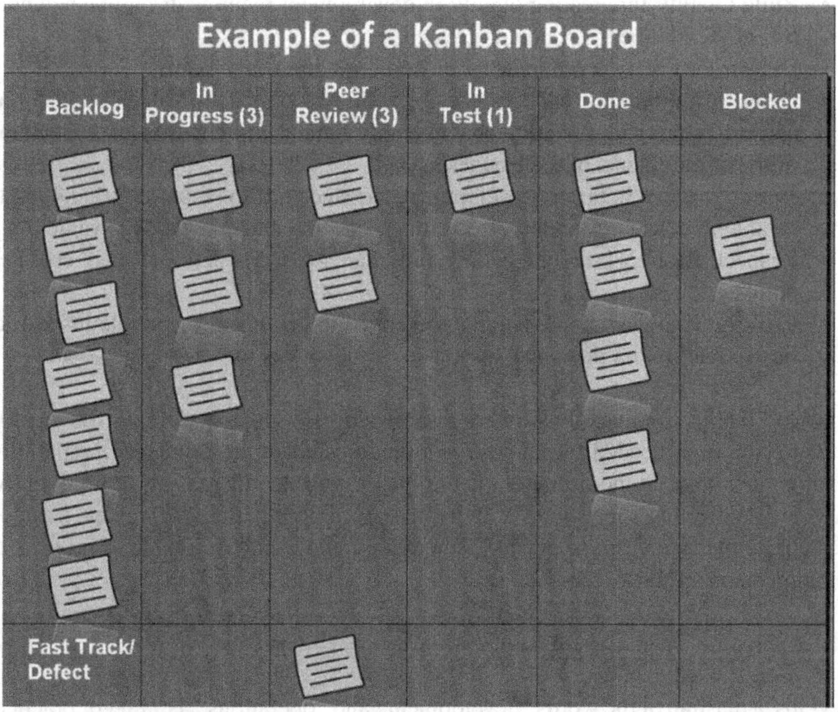

FIGURE 2.9   Example of Kanban Board

Kanban teams write the work items onto cards, usually one item per card. The cards are placed on the board, typically divided into columns, that show each flow of the software production (e.g., In Progress, In Testing, Completed, etc.). As the development evolves, the cards contained on the board are moved to the proper column that represents its status, and when a new task comes into play, a new card is created. Work moves around the board as capacity permits, rather than being pushed into the process when requested.

The Kanban method requires transparency and real-time communication so that the members of a team can know exactly at what stage the development is and can see the status of the project at any time. Kanban has been effectively used to develop large software and software-intensive systems.

2.3.39 WHAT IS LEAN DEVELOPMENT?

Lean Development encompasses hardware, software, and hybrid systems. "Lean Software Development" is borrowed from "Lean Manufacturing", which is a set of seven principles adapted from the Toyota Production System. The seven principles are:

- Eliminate Waste: everything that does not bring effective value to the customer's project shall be deleted.
- Build Quality In: creating quality in development requires discipline and control of the number of residuals created.
- Deliver Fast: deliver value to the customer as soon as possible.
- Defer Commitments: encouraging the team not to focus too much on planning and anticipating ideas without having a prior and comprehensive understanding of the requirements of the business.
- Optimize the Whole: the development sequence has to be perfected enough to be able to delete errors in the code, in order to create a flow of true value.
- Respect People: communicating and managing conflicts are two essential points.
- Creating Knowledge: the team is motivated to document the whole infrastructure to later retain that value.

Applying these principles in software and systems development allows the team to decrease the time needed to deliver functionalities since the principles prepare the development team in the decision making process, hence increasing general motivation, while deleting superfluous activity, therefore saving time and money. Besides, these principles are easily scalable methodology and easily adaptable to projects of any dimension.

2.3.40 WHAT ARE DEVOPS PRACTICES?

DevOps, a mashup of the words "development" and "operations", is "a set of practices intended to reduce the time between committing a change to a system and the change being placed into normal production, while ensuring high quality" (Bass, Weber, &

Zhu, 2015). Concepts leading to DevOps far predate its origin, but it began to appear in the industry around 2007/2008. DevOps practices involve four main concerns:

- Getting changes quickly into production
- Using automated testing to find errors
- Reducing or eliminating errors occurring during deployment
- Finding and repairing system faults quickly.

Since it is well-known that the cost of fixing requirements errors late in the development process is much higher than finding them early in the process, addressing the DevOps concerns suggests that a strong Requirements Engineering (RE) program is needed. For example, ensuring that an effective means for capturing non-functional requirements (NFRs) is in place as these are particularly difficult to address late in the process.

2.3.41 WHAT IS SCALED AGILE FRAMEWORK?

Scaled Agile Framework (SAFe) begin to emerge around the same time as DevOps as an agile style of development that incorporates elements of lean–agile, systems thinking, DevOps, and more. SAFe is also based on principles that predate its origin significantly. One of the major differences between SAFe, however, and other agile development methodologies is that SAFe is designed to scale up to much larger teams and projects.

Like the Agile Manifesto, SAFe is based on ten principles which include taking a system thinking and value-based approach to managing project requirements with an iterative build-out of functionality in working non-throwaway prototypes (also found in DevOps). SAFe also adopts other principles from agile development such as sustainable pace, shared ownership, and keeping participants motivated in the process (though these terms are not precisely used).

SAFe defines several requirement types that capture functionality from high level to lower level, conceptional, and NFRs. The intention of these types is to allow for a greater level of generality early in the development process with idiosyncratic requirements not being defined until later in the project. SAFe is itself adaptive, for example, it can work with many of the requirements representations and practices previously described such as User Stories, Use Cases, Scrum, XP, and Kanban.

2.3.42 WHEN SHOULD AGILE METHODOLOGIES BE USED?

The use of agile software methodologies is becoming very popular in the industry, at least as evidenced by survey data. For example, in a Forrester/Dr. Dobbs developer survey, 35% of respondents reported using some kind of agile methodology (West, Grant, Gerush, & D'Silva, 2010). More recently, a 2020 survey on requirements engineering state of practices (Kassab & Laplante, 2022), found that 70% of software engineers surveyed used an agile methodology as software development life cycle for their project making agile the most popular SDLC (Figure 2.10). The usage

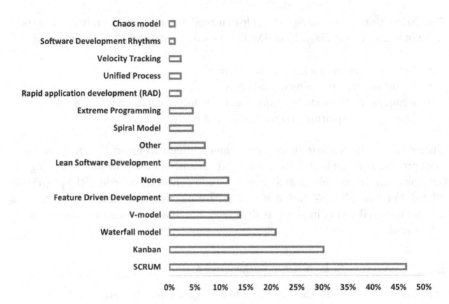

FIGURE 2.10 Software development life cycle employed

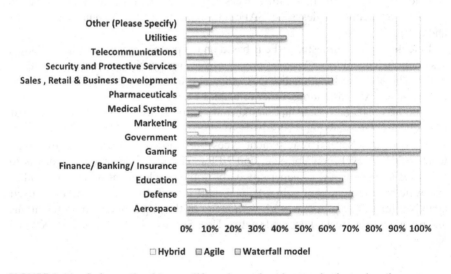

FIGURE 2.11 Software development life cycle employed across business domains

of agile is almost increased by more than 24% since 2013 when a similar survey was conducted (Kassab, Neill, & Laplante, 2014). The 2020 survey also found that agile methods were used more frequently than waterfall style development in every industry (Figure 2.11), and it observed a shift in the finance/banking/insurance, defense, and aerospace domains from 2013 when the waterfall was more popular than agile.

Stay Updated. For the up-to-date data from the state of practice survey: https://phil.laplante.io/requirements/updates/

But the question still arises, when should agile development methodologies be used? Boehm and Turner (2004) suggest that the way to answer the question is to look at the project along a continuum of five dimensions; size (in terms of the number of personnel involved), system criticality, personnel skill level, dynamism (anticipated number of system changes over some time interval), and organizational culture (whether the organization thrives on chaos or order) (Boehm & Turner, 2004). In Figure 2.12, as project characteristics tend away from the center of the diagram, then the likelihood of succeeding using agile methodologies decreases.

Therefore, projects assessed in the innermost circle are likely candidates for agile approaches, those in the second circle (but not in the inner circle) are marginal, and those outside of the second circle are not good candidates for agile approaches.

2.3.43 Is There a Case to Be Made for Not Using Agile Methods?

Like all engineering solution approaches, there are situations when agile software engineering should be used and there are situations when it should not be used. But it is not always easy to make this distinction. Ongoing misuse or misunderstanding can cloud the decision as to when to use agile approaches.

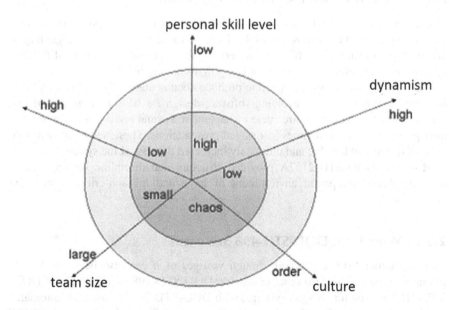

FIGURE 2.12 Balancing agility and discipline (Adapted from Boehm & Turner, 2004)

2.3.44 ALL OF THESE PROCESS MODELS LOOK RATHER SIMPLISTIC, ARTIFICIAL, OR TOO PRESCRIPTIVE. SHOULD THEY REALLY BE USED?

It has been suggested by Parnas and Clements (1986) that after the project has been completed, tested, and delivered, users can "cover their tracks" by modifying the documentation so that it appears that a deliberate methodology was used. For example, when the sequence of the waterfall model cannot be followed strictly, at least the documentation should suggest that it was followed in that sequence.

While this kind of practice might appear disingenuous, the benefit is that a traceable history is established between each program feature and the requirement driving that feature. This approach promotes a maintainable, robust, and reliable product and, in particular, one where decisions related to timing requirements are well-documented. It does indicate, however, that perhaps the process used was a reactive one and not part of a planned strategy.

2.4 SOFTWARE STANDARDS[2]

2.4.1 WHO PUBLISHES SOFTWARE STANDARDS?

Standardizing organizations such as ISO, ACM, IEEE, the U.S. Department of Defense (DOD), and others actively promote the development, use, and improvement of standards for software processes and inherent life cycle models. Even though many are interrelated and mutually influenced, the array of standards available can be confusing and even contradictory to the point of frustration.

2.4.2 WHAT IS THE DOD-STD-2167A STANDARD?

This extinct standard had a great deal of influence on the development of military software systems in the 1980s and 1990s. Because the U.S. DOD is the single largest procurer of software, the 2167A and waterfall culture pervades suppliers of military systems software even today, and so it is worth discussing briefly.

DOD-STD-2167A[3] was designed to produce documentation that achieves a high-integrity description of the evolving software design for baseline control and that serves as the foundation for life cycle management. Formal reviews were prescribed throughout but were sometimes just staged presentations. These audits often proved to be of questionable value and ultimately increased the cost of the system.

However, DOD-STD-2167A provided structure and discipline for the chaotic and complex development environment of large and mission-critical embedded applications.

2.4.3 WHAT IS THE DOD-STD-498 STANDARD?

This is another extinct standard, though vestiges of it can also be found widely throughout the defense and other industries. DOD-STD-498 was a merger of DOD-STD-2167A, used for weapon systems, with DOD-STD-7935A, used for automated information systems. Together, they formed a single software development standard for all of the organizations in the purview of the U.S. DOD. The purpose of developing this new standard, which was approved in 1994, was to resolve issues raised in the use of the old standards, particularly related to their incompatibility with modern software engineering practice.

The process model adopted in DOD-STD-498 was significantly different from 2167A. The former standard explicitly imposed a waterfall model, whereas 498 provided for a development model that is compatible with all of the software life cycle models discussed previously, except the lightweight methodologies.

2.4.4   What Is the ISO 9000 Standard?

ISO (International Organization for Standardization) 9000 is a generic, worldwide standard for quality improvement. The standard, which collectively is described in five standards (ISO 9000 through ISO 9004), was designed to be applied in a wide variety of manufacturing environments. ISO 9001 through ISO 9004 apply to enterprises according to the scope of their activities. ISO 9004 and the ISO 9000-X family are documents that provide guidelines for specific applications domains.

2.4.5   Which Part of ISO 9000 Applies to Software?

For software development, currently ISO/IEC/IEEE 90003:2018 is the document of interest. ISO originally released the 9000-3 quality guidelines to help organizations apply the ISO 9001 requirements to computer software. ISO 90003 is essentially an expanded version of ISO 9001 with added narrative to encompass software.

2.4.6   Who Uses This Standard?

ISO/IEC/IEEE 90003 is widely adopted in Europe, and an increasing number of U.S. and Asian companies have adopted it as well.

2.4.7   What Is in the ISO 90003 Standard?

The ISO standards are process-oriented, common sense practices that help companies create a quality environment. The principal areas of quality focus are:

Management responsibility
Quality system requirements
Contract review requirements
Product design requirements
Document and data control
Purchasing requirements
Customer supplied products
Product identification and traceability
Process control requirements
Inspection and testing
Control of inspection, measuring, and test equipment

Inspection and test status
Control of nonconforming products
Corrective and preventive actions
Handling, storage, and delivery
Control of quality records
Internal quality audit requirements
Training requirements
Servicing requirements
Statistical techniques.

Paying particular attention to some of these areas, such as inspection and testing, design control, and product traceability (through a "rational design process"), can increase the quality of a software product.

2.4.8   HOW SPECIFIC IS ISO 90003 FOR SOFTWARE?

Unfortunately, the standard is very general and provides little specific process guidance. For example, Figure 2.13 illustrates ISO 90003: 4.4 Software development and design. While these recommendations are helpful as a checklist, they provide very little in terms of a process that can be used.

While a number of metrics have been available to add some rigor to this somewhat generic standard, in order to achieve certification under the ISO standard, significant paper trails and overhead are required.

2.4.9   WHAT IS ISO/IEC STANDARD 12207?

ISO 12207: Standard for Information Technology — Software Life Cycle Processes, describes five "primary processes"—acquisition, supply, development, maintenance,

ISO 9000-3	4.4 Software development and design
4.4.1 General	Develop and document procedures to control the product design and development process. These procedures must ensure that all requirements are being met.
Software development	Control your software development project and make sure that it is executed in a disciplined manner.

      • Use one or more life cycle models to help organize your software development project.
      • Develop and document your software development procedures. These procedures should ensure that:
         • Software products meet all requirements.
         • Software development follows your:
            • Quality plan.
            • Development plan.

FIGURE 2.13   Excerpt from ISO 90003: 4.4 Software development and design

and operation. ISO 12207 divides the five processes into "activities," and the activities into "tasks," while placing requirements upon their execution. It also specifies eight "supporting processes"—documentation, configuration management, quality assurance, verification, validation, joint review, audit, and problem resolution—as well as four "organizational processes"—management, infrastructure, improvement, and training.

The ISO standard intends for organizations to tailor these processes to fit the scope of their particular projects by deleting all inapplicable activities, and it defines ISO 12207 compliance as the performance of those processes, activities, and tasks selected by tailoring.

ISO 12207 provides a structure of processes using mutually accepted terminology, rather than dictating a particular life cycle model or software development method. Because it is a relatively high-level document, ISO 12207 does not specify the details of how to perform the activities and tasks comprising the processes. Nor does it prescribe the name, format, or content of documentation. Therefore, organizations seeking to apply ISO 12207 need to use additional standards or procedures that specify those details.

The IEEE recognizes this standard with the equivalent numbering "IEEE/ EIA 12207.0-1996 IEEE/EIA Standard Industry Implementation of International Standard ISO/IEC12207:1995 and (ISO/IEC 12207) Standard for Information Technology — Software Life Cycle Processes."

NOTES

1   The term *Artifact* is sometimes used to mean software or software-related products such as documentation.
2   Some of the following discussion is adapted from the excellent text on software standards by Wang and King (2000).
3   DOD standards are sometimes referred to as "MIL-STD," for "military standard." So "DOD- STD-498" is equivalent to "MIL-STD-498."

FURTHER READING

Ambler, S. (2006). "Are you Agile or Fragile?" presentation, www.whysmalltalk.com/Smalltalk_Solutions/ss2003/pdf/ambler.pdf last accessed April 10.

Babcock, C. (1985). New Jersey Motorists in Software Jam, *Computerworld*, September, *30*, 1–6.

Bass, L., Clements, P., & Kazman, R. (2021). *Software architecture in practice*. 4th edition. Addison-Wesley Professional.

Bass, L., Weber, I., & Zhu, L. (2015). *DevOps: A Software Architect's Perspective*. Addison-Wesley Professional, Boston, MA, USA

Beck, K. (1999). Embracing change with extreme programming. *Computer, 32*(10), 70–77.

Beck, K., & Andres, C. (2004). *Extreme Programming Explained: Embrace Change*. 2nd edition. Addison-Wesley Professional.

Boehm, B.W. (1988). A spiral model of software development and enhancement. *Computer, 21*(5), 61–72.

Boehm, B.W., Brown, J.R., & Lipow, M. (1976). Quantitative Evaluation of Software Quality, In *proceeding of the 2nd International Conference on Software Engineering*, San Francisco, CA, Long Branch, CA: IEEE Computer Society, 592–605.

Boehm, B.W., & In, H. (1996). Identifying Quality-Requirement Conflicts. *IEEE Software*, IEEE Computer Society Press, 25–35.

Boehm, B.W., & Turner, R. (2004). *Balancing Agility and Discipline: A Guide to the Perplexed*. Addison-Wesley, Boston, MA.

Breitman, K.K., Leite J.C.S.P., & Finkelstein, A. (1999). The World's Stage: A Survey on Requirements Engineering Using a Real-Life Case Study. *Journal of the Brazilian Computer Society, 1*(6), 13–37.

Brooks, F.P. (1995). *The Mythical Man-Month*, 20th Anniversary Edition, Addison-Wesley, Boston, MA.

Chung, L., Nixon, B.A., Yu, E., & Mylopoulos, J. (2000). *Nonfunctional Requirements in Software Engineering*. Kluwer Academic Publishing.

IEEE (Institute of Electrical and Electronic Engineers) (1999). *IEEE 1473-1999, IEEE Standard for Communications Protocol Aboard Trains*.

ISO 90003 STANDARD www.iso.org/standard/74348.html

ISO 25000 STANDARDS https://iso25000.com/index.php/en/iso-25000-standards

Kassab, M. (2009). *Non-functional requirements: modeling and assessment*. VDM Verlag.

Kassab, M., DeFranco, J., & Laplante, P. (2020). A systematic literature review on Internet of things in education: Benefits and challenges. *Journal of Computer Assisted Learning, 36*(2), 115–127.

Kassab, M., DeFranco, J., & Voas, J. (2018). Smarter education. *IT Professional, 20*(5), 20–24.

Kassab, M. & Laplante, P.A. (2022). The Current and Evolving Landscape of Requirements Engineering State of Practice. *IEEE Software*. DOI Bookmark: 10.1109/MS. 2022.3147692

Kassab, M., Neill, C., & Laplante, P.A. (2014). State of practice in requirements engineering: Contemporary data. *Innovations in Systems and Software Engineering, 10*(4): 235–241.

Laplante, P.A. (2004). *Software Engineering for Image Processing Systems*. CRC Press, Boca Raton, FL.

Laplante, P.A., Kassab, M., Laplante, N.L., & Voas, J.M. (2017). Building caring healthcare systems in the Internet of Things. *IEEE Systems Journal, 12*(3), 3030–3037.

Martin, R.C. (2002). *Agile Software Development: Principles, Patterns, and Practices*. Prentice-Hall, Englewood Cliffs, NJ.

Neill, C.J. & Laplante, P.A. (2003). Requirements engineering: the state of the practice. *Software, 20*(6), 40–46.

Nerur, S., Mahapatra, R., & Mangalaraj, G. (2005). Challenges of migrating to agile methodologies. *Commun. ACM, 48*(5), 73–78.

Nord, R.L. & Tomayko, J.E. (2006). Software architecture-centric methods and agile development. *Software, 23*(2), 47–53.

O'Neil, F. (2015, July). Target data breach: applying user-centered design principles to data breach notifications. In Proceedings of the 33rd Annual International Conference on the Design of Communication, 1–8.

Parnas, D.L. & Clements, P.C. (1986). A rational design process: how and why to fake it. *IEEE Trans. Software Eng., 12*(2), 251–257.

Reifer, D. (2002). How good are agile methods? *Software, 19*(4), 16–18.

Royce, W. (1998). *Software Project Management: A Unified Framework*. Addison-Wesley, Boston, MA.

Schwaber, K., & Beedle, M. (2001). *Agile Software Development with SCRUM*. Prentice Hall. Upper Saddle River, NJ.

Siemens Warns of Possible Hearing Damage in Some Cell Phones. (2004). www.consumeraffa irs.com/news04/siemens_mobile.html, last visited on Aug, 4th, 2009.

The Rules and Practices of Extreme Programming, www.extremeprogramming.org/rules.html, accessed September 14, 2006.

Theuerkorn, F. (2005). *Lightweight Enterprise Architectures*. Auerbach Publications, Boca Raton, FL.

Tucker, A.B., Jr. (Editor-in-Chief) (1996). *The Computer Science and Engineering Handbook*. CRC Press, Boca Raton, FL.

Wang, Y.W., & King, G. (2000). *Software Engineering Processes: Principles and Applications*. CRC Press, Boca Raton, FL.

West, D., Grant, T., Gerush, M., & D'Silva, D. (2010). *Agile Development: Mainstream Adoption Has Changed Agility*. Forrester Research, Cambridge, MA.

3 Software Requirements Engineering

OUTLINE

- Requirements engineering concepts
- Requirements elicitation
- Requirements representation and documentation
- Requirements validation and verification
- Requirements management and tools
- Requirements in an agile environment

3.1 INTRODUCTION

Requirements engineering is the branch of software engineering concerned with the real-world goals for, functions of, and constraints on software systems. It is also concerned with the relationship of these factors to precise specifications of software behavior, and to their evolution over time and across software families (Zave, 1997). Different approaches to requirements engineering exist, some more complete than others. Whatever the approach taken, it is crucial that there is a well-defined methodology, and that documentation exists for each stage. This chapter incorporates a discussion of these aspects of requirements engineering for software development.

3.2 REQUIREMENTS ENGINEERING CONCEPTS

3.2.1 WHAT IS A "REQUIREMENT"?

Part of the challenge in requirements engineering has to do with an understanding of what a "requirement" really is. Requirements can range from high-level, abstract statements, and back-of-the-napkin sketches to formal (mathematically rigorous) specifications. These varying representations occur because stakeholders have needs at different levels, hence, depend on different abstraction representations.

A fundamental challenge for the requirements engineer is recognizing that customers often confuse requirements, features, and goals (and engineers sometimes do too).

While goals are high-level objectives of a business, organization, or system, a requirement specifies how a goal should be accomplished by a proposed system. On the other hand, a feature is a set of logically related requirements that allows the user to satisfy the goal.

DOI: 10.1201/9781003218647-4

So, to some stakeholders of a Home Automation System (HAMS), the goal is to create a system that allows for a business expansion by entering new and emerging geographic markets. This goal can be satisfied by implementing some features. One of these features will be by supporting international languages. Then, there will be a set of specific requirements to describe the mechanism of creating a system that can be customized to support different international languages. So, the requirements tend to be more granular than a feature and tend to be written with the implementation in mind.

To treat a goal as a requirement is to invite trouble because the achievement of the goal will be difficult to prove. In addition, goals evolve as stakeholders change their minds and refine and operationalize goals into behavioral requirements.

3.2.2 What Is the Requirements Specification?

This is the set of activities designed to capture behavioral and nonbehavioral aspects of the system in the software requirements specifications (SRS) document. The goal of the SRS activity, and the resultant documentation, is to provide a complete description of the system's behavior without describing the internal structure. This aspect is easily stated, but difficult to achieve, particularly in those systems where temporal behavior must be described.

3.2.3 What Are the Different Types of Requirements?

One common taxonomy for requirements specifications focuses on the type of requirement from the following list of possibilities:

- Functional requirements (FRs)
- Non-functional requirements (NFRs)
- Domain requirements.

3.2.4 What Are the "Functional Requirements"?

Functional requirements (FRs) describe the services the system should provide and how the system will react to its inputs. In addition, the functional requirements need to explicitly state certain behaviors that the system should not do (more on this later). Functional requirements can be high-level and general (in which case they are user requirements in the sense that was explained previously) or they can be detailed, expressing inputs, outputs, exceptions, and so on (in which case they are the system requirements described before).

There are many forms of representation for functional requirements, from natural language (in our case, the English language), visual models, and the more rigorous formal methods. An example of a functional requirement from the HAMS is:

When deployed in the U.S., the system shall convert measures received from external sensors from Metric to Imperial system.

3.2.5 What Are "Non-Functional Requirements"?

While software systems are characterized by functional behavior (what the system does), they are also characterized by their non-functional behavior (how the system behaves concerning some observable attributes like reliability, reusability, maintainability, etc.). In the software/system marketplace, in which functionally equivalent products compete for the same customer, non-functional requirements (NFRs) become more important in distinguishing between the competing products. They basically deal with issues like:

- Security
- Reliability
- Maintainability
- Performance
- Usability
- Testability
- Interoperability
- Constraints (e.g., Design/Implementation Constraints, Financial Constraints, Resources Constraints, etc.).

Consider the following NFRs from the HAMS:

The system shall convert measures received from external sensors from Metric to Imperial system in less than 1 second. (Performance requirement).
   The system shall perform the unit conversion correctly 99.99% of the time. (Reliability requirement).

In the last chapter, we discussed some of the most popular quality requirements.

3.2.6 What Are "Domain Requirements"?

Domain requirements are derived from the application domain. These types of requirements may consist of new functional requirements or constraints on existing functional requirements, or they may specify how particular computations must be performed.
   In the HAMS, various domain realities create requirements. There are industry standards (we wouldn't want the new system to underperform vs. other systems). There are constraints imposed by existing hardware available (e.g., existing sensor types). And there may be constraints on performance mandated by licensing authority (e.g., a constraint on the time interval the system has to spend till generating an alarm notification in case of a fire).

3.2.7 When Does Requirements Engineering Start?

Ideally, the requirements engineering process begins with a study activity, which leads to a feasibility report. It is possible that the feasibility study may lead to a decision not to continue with the development of the software product. If the feasibility study suggests that the product should be developed, then requirements analysis can begin.

3.2.8   What Are the Core Activities of Requirements Engineering?

The requirements engineer is responsible for a number of activities. These include:

- Requirements elicitation/discovery
- Requirements representation and documentation
- Requirements verification and validation
- Requirements management.

We will review each of these activities in the following sections.

3.3   REQUIREMENTS ELICITATION/DISCOVERY

3.3.1   What Is "Requirements Elicitation"?

Requirements elicitation/discovery involves uncovering what the customer needs and wants. But elicitation is not like harvesting low-hanging fruit from a tree. While some requirements will be obvious (e.g., the HAMS shall generate alarm notifications in case of a fire), many requirements will need to be extricated from the customer through well-defined approaches. This aspect of requirements engineering also involves discovering who the stakeholders are; for example, are there any hidden stakeholders? Elicitation also involves determining the NFRs, which are often overlooked.

3.3.2   What Are the Obstacles That May Challenge the Requirements Elicitation?

The following are common obstacles in the requirements elicitation process:

- New project domain—when requirements engineer doesn't possess enough knowledge on the industry or the developed solution. Engaging a domain expert can help alleviate this problem.
- Unclear project vision—when stakeholders don't have a clear understanding of what functionality their system needs. A clear mission statement or conops[1] can be very helpful to avoid this problem.
- Limited access to documentation—when the requirements engineer can't access documentation or when evaluating the current state of the project takes too much time. Consistent and disciplined use of a good document/change management system is important to combat this problem.
- Focus on the solution instead of requirements—when the customer focuses more on solutions or architectural tactics instead of the requirements themselves. Active expectation management is very important in this regard.

- Fixation of specific functionalities—when stakeholders insist on designing certain features because they believe they will benefit their business even when they will not. Again, expectation management and refocus on the mission statement is the key to addressing this issue.
- Contradictory requirements—when a project includes a wide range of stakeholders, their requirements may contradict each other.

3.3.3 How Should We Prepare for the Requirements Elicitation Process?

Identifying all customers and stakeholders is the first step in preparing for requirements elicitation. But stakeholder groups, and especially customers, can be nonhomogeneous, and therefore you need to treat each subgroup differently. For example, the different subclasses of users for the HAMS include:

- Facility manager
- System administrator
- Field engineer
- Public safety authority.

Each of these subgroups of users has different desiderata[2] and these need to be determined. The process, then, to prepare for elicitation is:

- Identify all customers and stakeholders.
- Partition customers and other stakeholders groups into classes according to interests, scope, authorization, or other discriminating factors (some classes may need multiple levels of partitioning).
- Select a champion or representative group for each user class and stakeholder group.
- Select the appropriate technique(s) to solicit initial inputs from each class or stakeholder group.

3.3.4 What Are the Common Requirements Elicitation Techniques?

There are many general techniques that you can choose to conduct requirements elicitation to overcome the above challenges, and you will probably need to use more than one, and likely different ones for different classes of users/stakeholders. Common elicitation techniques include:

- Brainstorming
- Card sorting
- Crowdsourcing
- Designer as apprentice
- Domain analysis
- Ethnographic observation
- Goal-based approaches
- Questionnaires

- Introspection
- Joint application development (JAD)
- Quality function deployment (QFD)
- Viewpoints.

Comprehensive discussions on each of these techniques and more are provided in Chapter 3 of the book *Requirements Engineering for Software and Systems* (Laplante & Kassab, 2022). We will provide here a brief overview on each of these techniques.

	Quick access to a summary of the elicitation technique: https://phil.laplante.io/requirements/elicit/summary.php

3.3.4.1   What Is the "Brainstorming" Technique?

Brainstorming consists of informal sessions with customers and other stakeholders to generate overarching goals for the systems. Brainstorming can be formalized to include a set agenda, minute taking, and the use of formal structures. But the formality of a brainstorming meeting is probably inversely proportional to the creative level exhibited at the meeting. These kinds of meetings probably should be informal, even spontaneous, with the only structure embodying some recording of any major discoveries.

3.3.4.2   What Is the "Card Sorting" Technique?

This technique involves having stakeholders complete a set of cards that includes key information about functionality for the system/software product. It is also a good idea for the stakeholders/customers to include a ranking and rationale for each of the functionalities.

3.3.4.3   What Is the "Crowdsourcing" Technique?

Crowdsourcing is a business model that harnesses the power of a large and diverse number of people to contribute knowledge and solve problems. The term crowdsourcing is a combination of crowd and outsourcing and was coined in 2006 by *Wired* magazine author Jeff Howe in his article *The Rise of Crowdsourcing* (Howe, 2006).

Social media (e.g., LinkedIn, Facebook, Twitter, etc.) provide the best-developed mechanisms to efficiently get feedback from crowds. The more efficient the communication mechanism to the crowd, the more people can be included in the process. A typical process to capture the requirements via crowdsourcing include the following steps:

- Choosing the potential crowd to be targeted
- Choosing the proper social media tool that is both popular with that crowd and that will allow the requirements engineer to ask questions of the crowd

- Creating a community with the crowd (perhaps by sharing useful information and news that's relevant to the crowd)
- Once the "crowdsourced elicitation process" is set up, it comes down to asking the right questions when opportunities arise. Asking questions to the crowd will be only efficient if the right questions are asked. That's where expertise comes in, along with other requirements elicitation discussed in this chapter. Since most people don't know what they want, it is the requirements engineer's job to present the right questions in an effective manner in order to trigger useful insights from the crowd. The same types of questions that don't work in one-on-one elicitation will not work in crowd elicitation either (see the questionnaires technique).

The elicitation process via crowdsourcing is often carried on with the support of tools. Some developed crowd-based tools for requirements elicitation include: CrowdREquire (Adepetu, Khaja, Al Abd, Al Zaabi, & Svetinovic, 2012), Refine (Snijders et al., 2015), StakeRare (Lim & Finkelstein, 2011), and Requirements Bazaar (Renzel & Klamma, 2014).

3.3.4.4 What Is the "Designer as Apprentice" Technique?

Designer as apprentice is a requirements discovery technique in which the requirements engineer "looks over the shoulder" of the customer in order to learn enough about the customer's work to understand their needs. The relationship between customer and designer is like that between a master craftsman and apprentice. That is, the apprentice learns a skill from the master just as we want the requirements engineer (the designer) to learn about the customer's work from the customer. The apprentice is there to learn whatever the master knows. Therefore, the designer (requirements engineer/apprentice) will guide the customer (master) in talking about and demonstrating the work, which often leads to discovery of subtle, unarticulated details.

3.3.4.5 What Is the "Domain Analysis" Technique?

Domain analysis involves any general approach to assessing the "landscape" of related and competing applications to the system being designed. Such an approach can be useful in identifying essential functionality and, later, missing functionality. Domain analysis can also be used later for identifying reusable components (such as open-source software elements that can be incorporated into the final design). The QFD elicitation approach explicitly incorporates domain analysis, and we will discuss this technique shortly.

3.3.4.6 What Is the "Ethnographic Observation" Technique?

Ethnographic observation refers to any technique in which observation of indirect and direct factors inform the work of the requirements engineer. Ethnographic observation is a technique borrowed from social science in which observations of human activity and the environment in which the work occurs are used to inform the scientist in the study of some phenomenon. In the strictest sense, ethnographic observation involves long periods of observation (hence, an objection to its use as a requirements elicitation technique).

3.3.4.7 What Is the "Goal-Based Approaches" Technique?

Goal-based approaches comprise any elicitation techniques in which requirements are recognized to emanate from the mission statement, through a set of goals that lead to requirements. That is, looking at the mission statement, a set of goals that fulfill that mission is generated. These goals may be subdivided one or more times to obtain lower-level goals. Then, the lower-level goals are branched out into specific high-level requirements. Finally, the high-level requirements are used to generate lower-level ones.

For example, consider the business goal of HAMS statement:

- To create a system that will ensure increased profitability.

The following goals might be considered to fulfill this business goal:

- Goal 1: The system will allow expansion by entering new and emerging geographic markets.
- Goal 2: The system will support opening new sales channels in the form of value-added resellers (VARs).

These goals can then be decomposed into requirements using a structured approach such as goal question metric (GQM). GQM is an important technique used in many aspects of systems engineering such as requirements engineering, architectural design, systems design, and project management. GQM incorporates three steps: state the system's objectives or goals; derive from each goal the questions that must be answered to determine if the goal is being met; decide what must be measured in order to be able to answer the questions (Basili & Weiss, 1984).

For example, in the case of the HAMS, consider goal 2. It can be refined into a set of requirements:

- R1. The system shall support hardware devices from different manufacturers.
- R2. The system shall support conversions of nonstandard units used by the different hardware devices.

Here is the related question regarding R2: "what is the latency constraint for the "units" conversion?" This question suggests a requirement of the form:

When deployed in the U.S., the system shall convert measures received from external sensors from Metric to Imperial system in less than 1 second.

The associated metric for this requirement, then, is simply the latency constraint.

3.3.4.8 What Is the "Questionnaires" Technique?

Requirements engineers often use questionnaires and other survey instruments to reach large groups of stakeholders. Surveys are generally used at the early stages of the elicitation process to quickly define the scope boundaries.

Survey questions of any type can be used. For example, questions can be closed (e.g., multiple-choice, true-false) or open-ended—involving free-form responses.

Closed questions have the advantage of easier coding for analysis, and they help to bind the scope of the system. Open questions allow for more freedom and innovation but can be harder to analyze and can encourage scope creep.

Questionnaires can be conducted through interviews. Interviews involve in-person communication between two individual stakeholders or a small group of stakeholders (sometimes called a focus group).

3.3.4.9   What Is the "Introspection" Technique?

When a requirements engineer develops requirements based on what he thinks the customer wants, then he is conducting the process of introspection. In essence, the requirements engineer puts himself in the place of the customer and asks "if I were the customer I would want the system to do this …"

An introspective approach is useful when the requirements engineer's domain knowledge far exceeds that of the customer.

3.3.4.10   What Is the "Joint Application Development" Technique?

Joint application device (JAD) involves highly structured group meetings (sometimes called "mini-retreats") with system users, system owners, and analysts focused on a specific set of problems for an extended period of time. These meetings occur 4–8 hours per day and over a period lasting 1 day to a couple of weeks. JAD has even been adapted for multisite implementation when participants are not co-located (Cleland-Huang & Laurent, 2014). While traditionally associated with large, government systems projects, the technique can be used in industrial settings on systems of all sizes.

JAD and JAD-like techniques are commonly used in systems planning and systems analysis activities to obtain group consensus on problems, objectives, and requirements. Specifically, the requirements engineer can use JAD sessions for the concept of operation definition, system goal definition, requirements elicitation, requirements analysis, requirements document review, and more.

Planning for a JAD review or audit session involves three steps:

1. Selecting participants
2. Preparing the agenda
3. Selecting a location.

Great care must be taken in preparing each of these steps. Reviews and audits may include some or all of the following participants:

- Sponsors (e.g., senior management)
- A team leader (facilitator, independent)
- Users and managers who have ownership of requirements and business rules
- Scribes (i.e., meeting minutes and note-takers)
- Engineering staff.

The sponsor, analysts, and managers select a leader. The leader may be in-house or a consultant. One or more scribes (note-takers) are selected, normally from the software development team. The analyst and managers must select individuals from the user community. These individuals should be knowledgeable and articulate in their business area.

Before planning a session, the analyst and sponsor must determine the scope of the project and set the high-level requirements and expectations of each session. The session leader must also ensure that the sponsor is willing to commit people, time, and other resources to the effort. The agenda depends greatly on the type of review to be conducted and should be constructed to allow for sufficient time. The agenda, code, and documentation must also be sent to all participants well in advance of the meeting so that they have sufficient time to review them, make comments, and prepare to ask questions.

The following are some rules for conducting software requirements, design audits, or code walkthroughs. The session leader must make every effort to ensure these practices are implemented.

- Stick to the agenda
- Stay on schedule (agenda topics are allotted specific time)
- Ensure that the scribe is able to take notes
- Avoid technical jargon (if the review involves nontechnical personnel)
- Resolve conflicts (try not to defer them)
- Encourage group consensus
- Encourage user and management participation without allowing individuals to dominate the session
- Keep the meeting impersonal
- Allow the meetings to take as long as necessary.

The end product of any review session is typically a formal written document providing a summary of the items (specifications, design changes, code changes, and action items) agreed upon during the session. The content and organization of the document obviously depend on the nature and objectives of the session. In the case of requirements elicitation, however, the main artifact could be a first draft of the SRS.

3.3.4.11   What Is the "Quality Function Deployment" Technique?

Quality function deployment (QFD) is a technique for discovering customer requirements and defining major quality assurance points to be used throughout the production phase. QFD provides a structure for ensuring that customers' desiderata are carefully heard, then directly translated into a company's internal technical requirements—from analysis through implementation to deployment. The basic idea of QFD is to construct relationship matrices between customer needs, technical requirements, priorities, and (if needed) competitor assessment. In essence, QFD incorporates card sorting, and laddering, and domain analysis.

Because these relationship matrices are often represented as the roof, ceiling, and sides of a house, QFD is sometimes referred to as the "house of quality" (Figure 3.1; Akao, 1990).

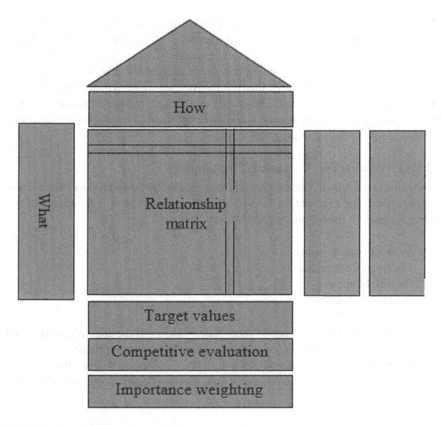

FIGURE 3.1 QFD's "house of quality" (Akao, 1990)

QFD was introduced by Yoji Akao in 1966 for use in manufacturing, heavy industry, and systems engineering. It has also been applied to software systems by IBM, DEC, HP, AT&T, Texas Instruments, and others.

When we refer to the "voice of the customer," we mean that the requirements engineer must empathically listen to customers to understand what they need from the product, as expressed by the customer in their words. The voice of the customer forms the basis for all analysis, design, and development activities, to ensure that products are not developed from only "the voice of the engineer." This approach embodies the essence of requirements elicitation.

The following requirements engineering process is prescribed by QFD:

- Identify stakeholders' attributes or requirements
- Identify technical features of the requirements
- Relate the requirements to the technical features
- Conduct an evaluation of competing products
- Evaluate technical features and specify a target value, for each feature
- Prioritize technical features for the development effort.

QFD uses a structured approach to competitive analysis. That is, a feature list is created from the union of all relevant features for competitive products. These features comprise the columns of a competition matrix. The rows represent the competing products and the corresponding cells are populated for those features included in each product. The matrix can be then used to formulate a starter set of requirements for the new or revised product. The matrix also helps to ensure that key features are not omitted from the new system and can contribute to improving the desirable quality of requirements completeness.

3.3.4.12   What Is the "Viewpoints" Technique?

Viewpoints are a way to organize information from the (point of view of) different constituencies. For example, in the HAMS, there are different perspectives of the system for each of the following stakeholders:

- Systems Administrators
- Maintenance engineers
- Facility Managers
- Regulatory agencies.

By recognizing the needs of each of these stakeholders and the contradictions raised by these viewpoints, conflicts can be reconciled using various approaches.

The actual viewpoints incorporate a variety of information from the business domain, process models, functional requirements specifications, organizational models, etc.

3.3.4.13   Which Combination of Requirements Elicitation Techniques Should Be Used?

There is scant research to provide guidance on selecting an appropriate mix of requirements elicitation techniques. One notable exception is the knowledge-based approach for the selection of requirements engineering techniques (KASRET), which guides users to select a combination of elicitation from a library (Eberlein & Jiang, 2011). The library of techniques and assignment algorithms are based on a literature review and survey of industrial and academic experts. The technique has not yet been widely used, however.

3.3.4.14   How May Artificial Intelligence Help in the Requirements Elicitation?

In the past few years, the thread of work on artificial intelligence (AI) for RE has made strides in rigorously investigating how general purpose AI tools can be tailored best for RE tasks (Dalpiaz & Niu, 2020). For example, recent research studies are investigating how human intervention in the requirement gathering processes can be reduced by using Speech Understanding Methodology techniques with the capability to "listen in" on a conversation and suitably collect stakeholders' declarations into a distinct vision (Spoletini & Ferrari, 2017; Sharma & Pandey, 2014).

Speech Understanding Methodology can be combined then with "Automatic Keywords Mapping", another AI technique being investigated that can enhance the requirements elicitations. Many requirements issues are related to the stakeholders being unable to depict their requirements properly, or the domain experts and developers not recognizing "observable" words that lead essentially to system requirements. These issues can be eliminated by automatically mapping every key-word spoken by each stakeholder. The earlier studies released a keyword mapping technique for designers so that they can recognize the keywords used by stakeholders to assist them in making ideal requirements (Sharma & Pandey, 2013).

Case-based reasoning is also being investigated for requirement elicitation, which can reduce the problem of natural language understanding as well as save the time of the requirement expert. There has also been recent research on the use of Machine Learning Algorithms to identify user preferences based on their sentiment (Li, Zhang, & Wang, 2018).

3.4   REQUIREMENTS REPRESENTATION AND DOCUMENTATION

3.4.1   What Is the Role of the SRS?

The SRS document is the official statement of what is required of the system developers. The SRS should include both a definition and a specification of requirements. However, the SRS is *not* a design document. As much as possible, it should be a set of *what* the system should do rather than *how* it should do it. Unfortunately, the SRS usually contains some design specifications, which has the tendency to hamstring the designers.

3.4.2   Who Uses the Requirements Documents?

A variety of stakeholders uses the software requirements throughout the software life cycle. Stakeholders include customers (these might be external customers or internal customers such as the marketing department), managers, developers, testers, and those who maintain the system. Each stakeholder has a different perspective on and uses for the SRS. Various stakeholders and their uses for the SRS are summarized in Table 3.1.

3.4.2.1   Within the SRS, Why Can't Requirements Just Be Communicated in English?

English, or any other natural language, is fraught with problems for requirements communication. These problems include lack of clarity and precision, mixing of functional and non-functional requirements, and requirements amalgamation, where several different requirements may be expressed together. Other problems with nat-ural languages include ambiguity, over-flexibility, and lack of modularization.

These shortcomings, however, do not mean that natural language is never used in an SRS document. Every clear SRS must have a great deal of narrative in clear and concise natural language. But when it comes to expressing complex behavior, it is

TABLE 3.1
Users of SRS Document

Stakeholder	Use
Customers	Express how their needs can be met. They continue to do this throughout the process as their perceptions of their own needs change.
Managers	Bid on the system and then control the software production process.
Developers	Create a software design that will meet the requirements.
Test Engineers	A basis for verifying that the system performs as required.
Maintenance Engineers	Understand what the system was intended to do as it evolves over time.

best to use formal or semi-formal methods, clear diagrams or tables, and narrative as needed to tie these elements together.

3.4.2.2   Then, What Are the Approaches for Requirements Representations?

Generally, there are three forms of requirements representation:

- Informal
- Semi-formal
- Formal.

Requirements specifications can strictly adhere to one or another of these approaches, but, usually, they contain elements of at least two of these approaches (informal and one other).

3.4.2.3   What Is Informal Representation?

Informal representation techniques cannot be completely transliterated into a rigorous mathematical notation. Informal techniques include natural language (i.e., human languages), flow diagrams, ad hoc diagrams, and most of the elements that you may be used to seeing (in SRS). Natural language is, in fact, the dominant form of requirements for projects in the industry. All SRS documents will have some informal elements. We can state this fact with confidence because even the most formal requirements specification documents have to use natural language, even if it is just to introduce a formal element. Kassab and Laplante (2022) found that 47% of respondents reported that requirements were expressed in terms of natural language, that is, informally (Figure 3.2). As natural language is notoriously imprecise, using natural language may present fundamental challenges in constructing software and systems. Some approaches bypass the challenges by using a constrained form of natural language, ensuring some degree of precision without going as far as the formal notations. Examples of these approaches include Stimulus (Jeannet & Gaucher, 2015), Relax (Whittle, Sawyer, Bencomo, Cheng, & Bruel, 2009), and Requirements Grammar (Scott & Cook, 2004).

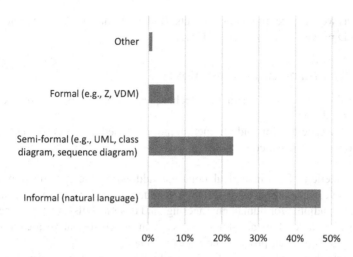

FIGURE 3.2   Requirement specification notation prevalence (From Kassab & Laplante, 2022)

 Stay Updated. For the up-to-date data from the RE state of practice survey: https://phil.laplante.io/requirements/updates/survey.php

3.4.3   WHAT IS SEMI-FORMAL REPRESENTATION?

Semi-formal representation techniques include those that, while appearing informal, have at least a partial formal basis. For example, many of the diagrams in the unified modeling language (UML) or systems modeling language (SysML) family of metamodeling languages, including the use case diagram, are semi-formal or can be formalized by adding certain semantic rules. UML and SysML are generally considered semi-formal modeling techniques; however, they can be made entirely formal with the addition of appropriate mechanisms. Other semi-formal techniques include KAOS (Dardenne, Van Lamsweerde, & Fickas, 1993), Rectify (Dassault Systems, 2016), and User Requirements Notations (URN) (Berenbach, Schneider, & Naughton, 2012).

3.4.4   WHAT IS UML/SYSML?

The UML is a set of modeling notations that are used in software and systems engineering requirements specification, design, systems analysis, and more. The SysML is an extension of UML dedicated to systems engineering and provides requirements diagrams, allowing users to express requirements in a textual representation and cover non-functional requirements. SysML diagrams can also express traceability links between requirements or between requirements and implementation elements, as well as other modeling artifacts. Both UML and SysML are quite similar. Appendix D provides an overview of UML.

3.4.5 WHAT IS FORMAL REPRESENTATION?

Formal methods involve mathematical techniques. To be precise, we define formal methods as follows:

A software specification and production method based on a precise mathematical syntax and semantics that comprises:

- A collection of mathematical notations addressing the specification, design, and development phases of software production, which can serve as a framework or adjunct for human engineering, and design skills and experience
- A well-founded logical inference system in which formal verification proofs and proofs of other properties can be formulated
- A methodological framework within which software may be developed from the specification in a formally verifiable manner Formal methods can be operational, denotational, or dual (hybrid)

It is clear from the definition that formal methods have a rigorous, mathematical basis. Formal methods differ from informal techniques, such as natural language, and informal diagrams like flowcharts. The latter cannot be completely transliterated into a rigorous mathematical notation. Of course, all requirements specifications will have informal elements, and there is nothing wrong with this fact. However, there are likely to be elements of the system specification that will benefit from formalization.

3.4.6 WHAT ADVANTAGE IS THERE IN APPLYING A LAYER OF POTENTIALLY COMPLEX MATHEMATICS TO THE ALREADY COMPLICATED PROBLEM OF BEHAVIORAL DESCRIPTION?

The answer is given by Forster, a giant of formal methods:

> One of the great insights of twentieth-century logic was that, in order to understand how formulae can bear the meanings they bear, we must first strip them of all those meanings so we can see the symbols as themselves ... [T]hen we can ascribe meanings to them in a systematic way ... That makes it possible to prove theorems about what sort of meanings can be borne by languages built out of those symbols.
>
> (Forster, 2003)

Formal methods are especially useful in requirements specification because they can lead to unambiguous, concise, correct, complete, and consistent specifications. But, as we have seen, achieving these qualities is difficult, and even with formalization correctness and completeness, it can be elusive. A thorough discussion of these issues along with a very detailed example can be found in Bowen, Hinchey, & Vassev (2010).

3.4.7 Which Notations to Select?

When making selections from different notations, Eberlein and Jiang (2011) suggest considering the following factors:

- Technical issues such as the maturity and effectiveness of the technique being considered
- The complexity of the techniques
- Social and cultural issues such as organization's opposition to change
- The level of education and knowledge of developers.

Certain characteristics of the software project, such as time and cost, constrain the complexity of the project, and the structure and competence of the software team.

3.4.8 What Is the Correct Format for an SRS?

There are many acceptable formats. For example, the IEEE 29148 standard (Figure 3.3) provides a general format for a requirements specification document. Other standards

```
1. Introduction
1.1 Purpose
1.2 Scope
1.3 Product overview
1.3.1 Product perspective
1.3.2 Product functions
1.3.3 User characteristics
1.3.4 Limitations
1.4 Definitions

2. References

3. Requirements
3.1 Functions
3.2 Performance requirements
3.3 Usability Requirements
3.4 Interface Requirements
3.5 Logical database requirements
3.6 Design constraints
3.7 Software system attributes
3.8 Supporting information

4. Verification
        (Parallel to subsections in Section 3)

5. Appendices
5.1 Assumptions and dependencies
5.2 Acronyms and abbreviations
```

FIGURE 3.3 Table of contents for an SRS document as recommended by IEEE Standard 29148 (From ISO/IEC/IEEE Standard 29148 Second edition, *Systems and Software Engineering—Life Cycle Processes—Requirements Engineering*, 2018)

bodies and professional organizations, such as the Project Management Institute (PMI), have standard requirements formats and templates. Most requirements management software will produce documents in customizable formats. But the "right" format depends on what the sponsor, customer, employer, and other stakeholders require.

That the SRS document should be easy to change is evident for the many reasons we have discussed so far. Furthermore, since the SRS document serves as a reference tool for maintenance, it should record forethought about the life cycle of the system, that is, to predict changes.

In terms of general organization, writing approach, and discourse, best practices include:

- Using consistent modeling approaches and techniques throughout the specification, for example, a top-down decomposition, structured, or object-oriented approaches
- Separating operational specification from descriptive behavior
- Using consistent levels of abstraction within models and conformance between levels of refinement across models
- Modeling non-functional requirements as a part of the specification models—in particular, timing properties
- Omitting hardware and software assignments in the specification (another aspect of design rather than the specification).

Following these rules will always lead to a better SRS document. Finally, IEEE 29148 describes certain desirable qualities for requirements specification documents and these will be discussed in the requirements validation & verification section below.

 Quick Access the latest version of the standard: https://phil. laplante.io/requirements/standard.php

3.4.9 WHAT IS THE RECOMMENDED FORMAT TO WRITE EACH REQUIREMENT?

Each requirement should be in a form that is clear, concise, and consistent in the use of language. A requirement can be written in the form of a natural language or some other form of language. If expressed in the form of a natural language, the statement should include a subject and a verb, together with other elements necessary to adequately express the information content of the requirement. Using consistent language assists users of the requirements documents (Hull, Jackson, & Dick, 2011) and renders analysis of the SRS document by software tools much easier.

A simplified standard requirement form is:

The [noun phrase] shall (not) [verb phrase].

Where [noun phrase] describes the main subject of the requirement, the shall (or shall not) statement indicates a required or prohibited behavior and [verb phrase] indicates the actions of the requirement.

Consider this requirement for the HAMS:

The system shall deny access for unauthorized users.

We assume that the terms "access," "deny", "unauthorized," and "user" have been defined somewhere in the SRS document.

It is not uncommon for requirements to be written in a nonstandard form. For example, the above requirement shown could be rewritten as:

Unauthorized users shall be denied access by the system.

Both versions of this requirement are equivalent but using active voice and avoiding using passive voice are recommended by the IEEE 29148. Representing requirements in the standard form is desirable for the reasons already mentioned.

Note that it is desirable to place measurable constraints on performance for functional requirements whenever possible, enriching the standard requirement form to:

[identifier] The [noun phrase] shall (not) [verb phrase] [constraint phrase].

As an example of a requirement that includes a measurable target, let's return to the above requirement, a realistic constraint might be incorporated to yield the following requirement:

The system shall deny access for unauthorized users within 1 second from attempted access.

Finally, "shall" is a command word or imperative frequently used in requirement specifications. Other command words and phrases that are frequently used include "should", "must", "will", "responsible for", and there are more. Wilson, Rosenberg, & Hyatt (1997) describe the subtle differences in these imperatives, which are summarized in Table 3.2.

3.5 REQUIREMENTS VALIDATION AND VERIFICATION

3.5.1 WHAT IS REQUIREMENTS RISK MANAGEMENT?

Poor requirements engineering practices have led to some notoriously spectacular failures. Bahill and Henderson (2005) analyzed several high-profile projects to determine if their failure was due to poor requirements development, requirements verification, or system validation. Their findings confirmed that some of the most notorious system failures were due to poor requirements engineering. For example, the HMS Titanic (1912), the IBM PCjr (1983), and the Mars Climate Orbiter (1999) all suffered from poor requirements design. The Apollo 13 (1970), Space Shuttle Challenger (1986), and Space Shuttle Columbia (2002) projects failed because of insufficient requirements verification. More recently, the rollout of the U.S. Healthcare.gov healthcare exchange site (2013) experienced numerous and very conspicuous problems due to a rushed implementation and failure to conduct proper requirements elicitations.

TABLE 3.2
Imperative Words and Phrases

Imperative	Most Common Use
Are applicable	To include, by reference, standards, or other documentation as an addition to the requirements being specified
Is required to	As an imperative in specifications statements written in the passive voice
Must	To establish performance requirements or constraints
Responsible for	In requirements documents that are written for systems whose architectures are predefined
Shall	To dictate the provision of a functional capability
Should	Not frequently used as an imperative in requirement specification statements
Will	To cite things that the operational or development environment are to provide to the capability being specified

Source: Wilson et al., 1997.

TABLE 3.3
History of Project Failures due to Poor Requirements Practices

System Name	Year	Requirements Process Failure
HMS Titanic	1912	Poor requirements design
Apollo-13	1970	Insufficient requirements verification
IBM PCjr	1983	Poor requirements design
Space Shuttle Challenger	1986	Insufficient requirements verification
Mars Climate Orbiter	1999	Poor requirements design
Space Shuttle Columbia	2002	Insufficient requirements verification
HealthCare.gov	2013	Poor requirements design
Boeing 737 Max	2019	Poor requirements design

More catastrophically, crashes of the Boeing 737 Max aircraft were eventually attributed to poor requirements design (Table 3.3).

Requirements can be inadequate in many ways including:

- Inaccurate or incomplete stakeholder identification
- Insufficient requirements validation
- Insufficient requirements verification
- Incomplete requirements
- Incorrect requirements
- Incorrectly ranked requirements
- Inconsistent requirements (Laplante, 2010).

Share your opinion. Do you have a real-world story of a project failure due to poor requirements practices?: https://phil.laplante. io/requirements/failures.php

Requirements risk management involves the proactive analysis, identification, monitoring, and mitigation of any factors that can threaten the integrity of the requirements engineering process. Requirements risk factors can be divided into two types: technical and management. Technical risk factors pertaining to the elicitation, analysis, agreement, and representation processes. The use of formal methods helps in improving the quality of requirements representation through mathematically rigorous approaches. Requirements management risk factors tend toward issues of expectation management and interpersonal relationships.

3.5.2 What Is the Difference between Validation and Verification?

Requirements validation and verification involve the review, analysis, and testing to ensure that a system complies with its requirements. Compliance pertains to both functional and non-functional requirements. But the foregoing definition does not readily distinguish between verification and validation, and is probably too long, in any case, to be inspirational or to serve as a mission statement for the requirements engineer. Boehm (1984) suggests the following to make the distinction between verification and validation:

- Requirements validation: "am I building the right product?"
- Requirements verification: "am I building the product right?"

In other words, validation involves fully understanding customer intent and verification involves satisfying the customer intent.

3.5.3 What Are the Techniques to Perform Validations and Verifications on Requirements?

V&V techniques may include some of the requirements elicitation techniques that were discussed. For example, group reviews/inspections, focus groups, viewpoint resolution, or task analysis (through user stories and use cases) can be used to simplify, combine, or eliminate requirements. In addition, we can use comparative product evaluations to uncover missing or unreasonable requirements and task analysis to uncover and simplify requirements. Systematic manual analysis of the requirements, test case generation (for testability and completeness), using an executable model of the system to check requirements, and automated consistency analysis may also be used. Wideband Delphi or the Analytical Hierarchy Process (AHP) can be used for requirements reconciliation, harmonization, and negotiation (more on this in Chapter 8). Finally, certain formal methods such as model checking and consistency analysis can be used for V&V.

3.5.4 Are There Any Tools to Support the Validation and Verification?

In the last two decades, there have been many attempts to build tools to support the formulation, documentation, and verification of natural language requirements. There are numerous tools that use natural language processing (NLP) to mitigate the problems and increase the quality of natural language requirements (see Table 3.4).

For example, Femmer et al. (2014) built a lightweight tool to detect "requirements smells," that is, various forms of phrases and word combinations that could be considered ambiguous, vague, incomplete, or problematic. Another tool, quality analyzer for requirements specification (QuARS), uses automated lexical and syntactic analyses to "identify requirements that are defective because of language usage (Lami, 2005)." Another natural language analysis tool, the NASA Automated Requirements Measurement (ARM), was retired in the 2010s. A faithful rendition of the ARM tool is maintained by Laplante and can be accessed here.

 Quick Acess to: NASA ARM Tool https://arm.laplante.io/

Other automated analysis tools require that the SRS conforms to standard templates and boilerplates. For example, the easy approach to requirements engineering (EARS) tool can find template violations. Similarly, the requirements template

TABLE 3.4
Tools That Focus on Finding Requirements Defects and Deviations

Tool Name	Year	Aim	Input	Automation
Circe	2006	Quality of requirements	Requirements document	Semi-automated
QuARS	2004	Quality of requirements	Requirements document	Automated
CRF Tool	2012	Uncertainty	Requirements document	Automated
Text2Test	-	Quality of use cases	Use cases	Unknown
AQUSA	2015	Quality of user stories	User stories	Automated
MaramaAI	2011	Quality of requirements	Requirements document	Not automated
SREE	2013	Ambiguity	Requirements document	Not available
Dowser	2008	Ambiguity	Requirements document	Not automated
RQA	2011	Quality of requirements	Requirements document	Automated
UIMA	2009	Use case model	Use case description	Not automated
Qualicen	2014	Quality of requirements	Requirements document	Automated
EARS	2010	Requirements Template Violations	Requirements document	Automated
RETA	2013	Requirements Template Violations	Requirements document	Automated
Planguage	2005	Quality of requirements	Requirements document	Automated

analyzer (RETA) can find template violations and also identify certain problematic requirements based on keywords (Arora, Sabetzadeh, Briand, & Zimmer, 2015).

Restricting the language that can be used in the specification document is another way to facilitate automated requirements V&V. One such tool, planning language (Planguage), uses a programming language-like syntax to structure requirements and then analyze them (Gilb, 2005).

Gervasi and Nuseibeh (2002) proposed a lightweight validation of natural language requirements (Circe) that can detect violations of quality characteristics in a more exact way by building logical models of the requirements specifications. Their approach, however, assumes that the specifications are written in certain patterns. This expectation is often not the case in the industry.

3.5.5 What Are the Standards for Validation and Verification?

There are various international standards for the processes and documentation involved in V&V of systems and software. Many of these have been sponsored or cosponsored by the IEEE. Whatever requirements V&V techniques are used, a software requirements V&V plan should always be written to accompany any major or critical software application. The IEEE Standard 1012-2012, "IEEE Standard for Software Verification and Validation," provides some guidelines to help prepare V&V plans. ISO/IEC/IEEE 29148 (we refer to it as IEEE 29148 for short) is a very important standard for requirements engineers; it "describes recommended approaches for the specification of software requirements."

3.5.6 What Are the Qualities That Must Exist in Requirements according to IEEE 29148?

IEEE 29148 specifies qualities for individual requirements and then for the set of requirements taken as a whole. First, the mandatory qualities for an individual requirement are that it must be:

- Singular: A requirement specifies a single behavior and has no conjunctions.
- Feasible: A requirement is feasible if it can be satisfied with current technology and cost constraints, that is, it is not a ridiculous requirement.
- Unambiguous: A requirement is unambiguous if it can have only one interpretation.
- Complete: A requirement is complete "to the extent that all of its parts are present and each part is fully developed".
- Consistent: The satisfaction of one requirement does not preclude the satisfaction of another, and, the SRS is in agreement with all other applicable documents and standards.
- Verifiable: The satisfaction of each requirement can be established using measurement or some other unambiguous means.
- Traceable: "The ability to describe and follow the life of a requirement in both a forwards and backward direction from inception throughout the entire system's life cycle, provides useful support mechanisms for managing requirement changes during the ongoing change process" (Gotel, 1995).

TABLE 3.5
A Sample Traceability Matrix

Requirement Identification	1.1	1.2	1.3	2.1	2.2	2.3	3.1	3.2
1.1		U	R					
1.2			U				R	U
1.3	R			R				
2.1			R		U			U
2.2								U
2.3		R		U				
3.1								R
3.2							R	

3.5.7 WHAT DOES A TRACEABILITY MATRIX LOOK LIKE?

One type of traceability matrix is shown in Table 3.5. Requirements identification numbers label both the rows and columns. An "R" is placed in a corresponding cell if the requirement in that row references the requirement in that column. A "U" corresponds to an actual use dependency between the two requirements.

Many other types of traceability matrices are possible, for example, relating requirements to test cases, requirements to applicable standard and regulations, and requirements to the primary stakeholder groups to which they relate.

3.6 REQUIREMENTS MANAGEMENT AND TOOLS

3.6.1 WHAT IS REQUIREMENTS MANAGEMENT?

Requirements management involves identifying, documenting, and tracking system requirements from inception through delivery. Inherent in this definition is the understanding of the true meaning of the requirements and the management of customer (and stakeholder) expectations throughout the system's lifecycle. A solid requirements management process is the key to a successful project.

3.6.2 WHAT ARE THE CHALLENGES TO REQUIREMENTS MANAGEMENT?

Hull et al. (2011) suggest that there are five main challenges to requirements management. The first is that very few organizations have a well-defined requirements management process, and thus, few people in those organizations have requirements management experience. The second challenge is the difficulty in distinguishing between user or stakeholder requirements and systems. The third problem is that organizations manage requirements differently, making the dissemination and transferability of experience and best practices difficult. The difficulties in progress monitoring are yet another problem. And finally, they suggest that managing changing requirements is a significant challenge. They suggest that establishing a well-defined requirements management process is key in addressing these issues.

TABLE 3.6
Common Commercial Tools Supporting RE Activities

Tool Name	Vendor	License	Capabilities
Jira	Atlassian	Commercial	Agile Support, Issue resolution management, project management, requirements management
Rational Doors	IBM	IBM EULA	Configurations management, requirements management, test management
Azure DevOps	Microsoft	Commercial	Agile support, issue resolution management, requirements management, test management
Jama Connect	Jama Software	Commercial	Project management, requirements management, test management
Doors Next (Jazz)	IBM	IBM EULA	Application lifecycle management, requirements management, test management
Aha!	Aha! Labs	Commercial	Product management, requirements management
Quality Center	Micro Focus	Commercial	Issue resolution management, requirements management, test management, project management
VersionOne	CollabNet	Commercial	Agile support, project management, requirements management

Source: Kassab & Laplante, 2022.

3.6.3 What Are the Existing Approaches to Reach Consensus among Different Stakeholders?

There are numerous approaches to consensus building. More on this will be provided in Chapter 8.

3.6.4 What Are Popular Tools to Support RE Activities?

Table 3.6 lists the top tools reported by the participants of the 2020 RE state of practices survey (Kassab & Laplante, 2022).

3.7 REQUIREMENTS ENGINEERING IN AN AGILE ENVIRONMENT

3.7.1 How Is RE Different in Agile?

The increasing adoption of agile practices in software development has been on the radar of the RE community in recent years (Wagner, 2018). This is mainly because "doing requirements" consumes time that could be rather spent on writing code, and business stakeholders in an agile environment often feel better if they think that the development team is "working" (that is, coding) on their problems (Orr, 2004).

While many conventional RE practices are important also in agile projects, as Savolainen, Kuusela, and Vilavaara (2010) observe, there are some major differences as well.

Paetsch, Eberlein, and Maurer (2003) analyzed commonalities and differences between traditional RE approaches and agile software development. Their findings include that one of the differences is related to customers involvement. Customers are constantly involved in requirements discovery and refinement in agile methods. All systems developers should be involved in the requirements engineering activity and each can, and should, have regular interaction with customers. In traditional approaches, the customer has less involvement once the requirements specification has been written and approved, and typically, the involvement is often not with the systems developers.

Another difference is in the timing of the requirements engineering activities. In traditional systems and software engineering, requirements are gathered, analyzed, refined, etc., at the front end of the process. In agile methods, requirements engineering is an ongoing activity; that is, requirements are refined and discovered with each system build. Even in spiral methodologies, where prototyping is used for requirements refinement, requirements engineering occurs much less so late in the development process.

A third noted difference in the agile approach to requirements engineering is much more invulnerable to changes throughout the process (remember, "embrace change") than in traditional software engineering.

The same findings were confirmed in a study by Ramesh, Cao, and Baskerville (2010) who examined how RE had been conducted in 16 organizations that were involved in agile software development.

Bose, Kurhekar, and Ghoshal (2008) suggested that while following agile practices may be effective in acquiring continuous feedback from customers, the limits of these practices on RE are not well-defined. They recommended that proper requirements management should be adopted under the agile umbrella to ensure proper traceability when the requirements are likely to be changed. The verification of early requirements representations was also proposed.

Cao and Ramesh (2008) studied 16 software development organizations that were using either XP or Scrum (or both) and uncovered seven requirements engineering practices that were agile in nature.

1. Face-to-face communications (overwritten specifications)
2. Iterative requirements engineering
3. Extreme prioritization (ongoing prioritization rather than once at the start, and prioritization is based primarily on business value)
4. Constant planning
5. Prototyping
6. Test-driven development
7. Reviews and tests.

Some of these practices can be found in non-agile development (e.g., test-driven development and prototyping), but these seven practices were consistent across all of the organizations studied.

As opposed to the fundamental software requirements specification, the fundamental artifact in agile methods is a stack of constantly evolving and refining requirements. These requirements are generally in the form of user stories. In any case, these requirements are generated by the customer and prioritized—the higher the level of detail in the requirement, the higher the priority. As new requirements are discovered, they are added to the stack and the stack is reshuffled in order to preserve prioritization.

There are no prohibitions on adding, changing, or removing requirements from the list at any time (which gives the customer tremendous freedom). Of course, once the system is built, or likely while is it is being built, the stack of user stories can be converted to a conventional software requirements specification for system maintenance and other conventional purposes.

Ambler (2007) suggests the following best practices for requirements engineering using agile methods. Many of the practices follow directly from the principles behind the Agile Manifesto:

- Have active stakeholder participation
- Use inclusive (stakeholder) models
- Take a breadth-first approach
- Model "storm" details (highly volatile requirements) just in time
- Implement requirements, do not document them
- Create platform-independent requirements to a point
- Remember that smaller is better
- Question traceability
- Explain the techniques
- Adopt stakeholder terminology
- Keep it fun
- Obtain management support
- Turn stakeholders into developers
- Treat requirements like a prioritized stack
- Have frequent, personal interaction
- Make frequent delivery of software
- Express requirements as features.

Ambler (2007) also suggests using such artifacts as CRCs, acceptance tests, business rule definitions, change cases, data flow diagrams, user interfaces, use cases, prototypes, features and scenarios, use cases diagrams, and user stories to model requirements. These elements can be added to the software requirements specification document along with the user stories.

For requirements elicitation, he suggests using interviews (both in-person and electronic), focus groups, JAD, legacy code analysis, ethnographic observation, domain analysis, and having the customer on-site at all times (Ambler, 2007).

Finally, a systematic literature review on agile requirements engineering practices and challenges (Inayat, Salim, Marczak, Daneva, & Shamshirband, 2015) identified 17 RE practices that we found to be adopted in agile software development:

- Face-to-face communication between team members and client representatives is a characteristic of agile RE.
- Customer involvement and interaction were declared the primary reasons for project success and limited failure.
- User stories are created as specifications of the customer requirements. User stories facilitate communication and better overall understanding among stakeholders.
- Iterative requirements, requirements unlike in traditional software development methods, emerge over time in agile methods.
- Requirement prioritization is part of each iteration in agile methods.
- Change management has proven to be a significant challenge for traditional approaches thus far.
- Cross-functional teams include members from different functional groups who have similar goals.
- Prototyping is perceived as a straightforward way to review requirements specifications with clients and to gain timely feedback before moving to subsequent iterations.
- Testing before coding means writing tests prior to writing functional codes for requirements.
- Requirements modeling is performed in agile software development methods, but it is different from RE models developed in traditional software development methods.
- Requirements management is performed by maintaining product backlog/feature lists and index cards.
- Review meetings and acceptance tests are the developed requirements and product backlogs that are constantly reviewed in meetings.
- Code refactoring is meant to revisit and modify developed code structure, improve on the structure, and accommodate changes.
- Shared conceptualization is a supporting concept to carry out RE activities related to gathering, clarifying, and evolving for agile methods.
- Pairing for requirements analysis is a practice that encourages the stakeholders to perform multiple roles as well.
- Retrospectives are the meetings held after the completion of an iteration.
- Continuous planning is a routine task for agile teams.

It is not necessary to adopt all of these practices nor are others forbidden. Each project, however, should be carefully evaluated to determine which of the above practices are appropriate and which others, if any, need to be added to that list.

3.7.2 How Is Requirements Engineering Performed in XP?

Requirements engineering in XP follows the model shown in Figure 3.4 where the stack of requirements in Ambler's model refers to user stories. In XP, user stories are managed and implemented as code via the "planning game."

The planning game in XP takes two forms: release and iteration planning. Release planning takes place after an initial set of user stories has been written.

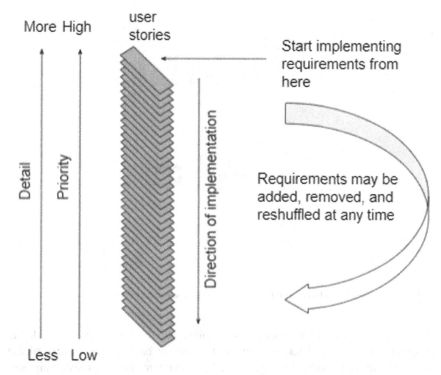

FIGURE 3.4 Agile requirements change management process (Adapted from www.agilem odeling.com/essays/agileRequirements.htm; Ambler, 2007)

This set of stories is used to develop the overall project plan and plan for iterations. The set is also used to decide the approximate schedule for each user story and overall project.

Iteration planning is a period of time in which a set of user stories and fixes to failed tests from previous iterations are implemented. Each iteration is 1–3 weeks in duration. Tracking the rate of implementation of user stories from previous iterations (which is called project velocity) helps to refine the development schedule.

Because requirements are constantly evolving during these processes, XP creator Kent Beck says that "in XP, requirements are a dialog, not a document" (Beck et al., 2001), although it is typical to convert the stack of user stories into a software requirements specification.

3.7.3 How Is Requirements Engineering Performed in Scrum?

In Scrum, the requirements stack shown in the model of Figure 3.4 is, as in XP, the evolving backlog of user stories. As in XP, these requirements are frozen at each iteration for development stability. In Scrum, each iteration takes about a month. To manage the changes in the stack, one person is given final authority for requirement prioritization (usually the product sponsor).

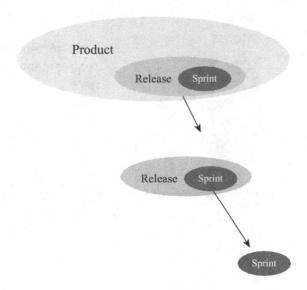

FIGURE 3.5   Backlog relationship between product, releases, and sprints

In Scrum, the requirements backlog is organized into three types: product, release, and sprint. The product backlog contains the release backlogs, and each release contains the sprint backlog. Figure 3.5 is a Venn diagram showing the containment relationship of the backlog items.

The product backlog acts as a repository for requirements targeted for release at some point. The requirements in the product backlog include low-, medium-, and high-level requirements.

The release backlog is a prioritized set of items drawn from the product backlog. The requirements in the release backlog may evolve so that they contain more details and low-level estimates.

Finally, the sprint backlog list is a set of release requirements that the team will complete (fully coded, tested, and documented) at the end of the sprint. These requirements have evolved to a very high level of detail, and hence, their priority is high.

Scrum has been adopted in many major corporations, with notable success. Some of the authors' students also use Scrum in courses. In these cases, it proves highly effective when there is little time for long requirements discovery processes.

3.7.4   How Are the User Stories Gathered?

User stories are a common unit of requirements in most agile methodologies. Each user story represents a feature desired by the customer. User stories gathering goes on throughout the project because user requirements keep on evolving throughout the project. Initial user stories are usually gathered in small offsite meetings. Stories can be generated either through goal-oriented (e.g., "let's discuss how a customer makes a purchase") approaches or through interactive (stream-of-consciousness) approaches.

Developing user stories is an iterative and interactive process. Formal requirements, use cases, and other artifacts are derived from the user stories by the software engineering team as needed.

3.7.5 How Are the User Stories Written?

User stories (a term coined by Beck) are written by the customer on index cards, though the process can be automated via wikis or other tools (Beck & Andres, 2004).

A user story consists of the following components:

- Title—this is a short handle for the story. A present tense verb in active voice is desirable in the title.
- Acceptance test—this is a unique identifier that will be the name of a method to test the story.
- Priority—this is based on the prioritization scheme adopted. Priority can be assigned based on "traditional" prioritization of importance or on the level of detail (higher priority is assigned to higher detail).
- Story points—this is the estimated time to implement the user story. This aspect makes user stories helpful for effort and cost estimation.
- Description—this is one to three sentences describing the story.

A sample layout for these elements on an index card is shown in Table 3.7.

An example user story for locking a user's account after failed attempts to access the HAMS with the wrong password is shown in Table 3.8. User stories should be understandable to the customers and each story should add value. Developers do not write user stories, users do. But stories need to be small enough that several can be completed per iteration. Stories should be independent (as much as possible); that is, a story should not refer back and forth to other stories. Finally, stories must be

TABLE 3.7
User Story Layout

Title		
Acceptance Test	Priority	Story Points
	Description	

TABLE 3.8
User Story: Lock Account after Failed Access Attempts

Title: Detect Security Threat		
Acceptance Test: detSecThrt	Priority: 1	Story Points: 3

When a user attempts to access the building manager's account with a wrong password 5 times in sequence, the account shall be blocked for 60 minutes. The security manager shall be sent an email stating that a potential threat has been detected.

testable—like any requirement, if it cannot be tested, it's not a requirement. The testability of each story must be considered by the development team.

3.7.6 How Are the User Stories Estimated?

To measure the user stories, agile teams allocate story points—arbitrary values to measure the effort required to complete that user story. These points can be allocated in many ways based on the team's preferences. In most cases, project managers define a story point complexity range as a Fibonacci series (e.g., 1, 2, 3, 5, 8). Another way is to pick a small reference story and estimate the other ones regarding that using the Delphi estimation technique known as "planning poker". The stories are rated and reviewed before entering the next phase of prioritization.

3.7.7 How Are the User Stories Prioritized?

User stories are prioritized before iteration. The user stories are prioritized in terms of ranking (i.e., ordered as first, second, and third) and also as a group (e.g., "high" priority, "low" and "medium"). The high-priority user stories are recorded in the product backlog and used as a guide to carry out the development work. Customers perform the prioritization task based on their understanding of the business value the user stories will bring using various techniques.

NOTES

1   Concept of Operations—a brief form of mission statement or definition of what the system is intended for and supposed to do.
2   Desiderata is Latin for "needs and desires".

FURTHER READING

Adepetu, A., Khaja, A.A., Al Abd, Y., Al Zaabi, A., & Svetinovic, D. (2012, March). CrowdREquire: A requirements engineering crowdsourcing platform. In 2012 AAAI Spring Symposium Series.

Akao, Y. (1990). *Quality Function Deployment: Integrating Customer Requirements into Product Design*. Productivity Press, Cambridge, MA.

Ambler, S. (2007). *Agile requirements change management*. www.agilemodeling.com/essays/changeManagement.htm (accessed December 2021).

Arora, C., Sabetzadeh, M., Briand, L., & Zimmer, F. (2015). Automated checking of conformance to requirements templates using natural language processing. *IEEE Transactions on Software Engineering, 41*(10), 944–996.

Bahill, T.A., & Henderson, S.J. (2005). Requirements development, verification, and validation exhibited in famous failures. *Systems Engineering 8*(1), 1–14.

Basili, V.R., & Weiss, D. (1984). A methodology for collecting valid software engineering data. *IEEE Transactions on Software Engineering 10*, 728–738.

Beck, K., & Andres, C. (2004). *Extreme Programming Explained: Embrace Change.* 2nd edition. Addison-Wesley Professional.

Beck, K., Beedle, M., van Bennekum, A., Cockburn, A., Cunningham, W., Fowler, M., Grenning, J., et al. (2001). *"Agile Manifesto" and "Principles Behind the Agile Manifesto."* http:// agilemanifesto.org/ (accessed December 2021).

Berander, P., Damm, L.O., Eriksson, J., Gorschek, T., Henningsson, K., Jönsson, P., ... & Wohlin, C. (2005). Software quality attributes and trade-offs. *Blekinge Institute of Technology*, *97*(98), 19.

Berenbach, B., Schneider, F., & Naughton, H. (2012, September). The use of a requirements modeling language for industrial applications. In *2012 20th IEEE International Requirements Engineering Conference (RE)* (pp. 285–290). IEEE.

Boehm, B.W. (1984). Verifying and validating software requirements and design specifications. *IEEE Software*, *1*(1), 75–88.

Bose, S., Kurhekar, M., & Ghoshal, J. (2008). Agile methodology in Requirements Engineering. *SETLabs Briefings Online*, 13–21.

Bowen, J.P., Hinchey, M.G., & Vassev, E. (2010). Formal requirements specification. In P. Laplante (Ed.), *Encyclopedia of Software Engineering*, pp. 321–332. Taylor & Francis. Boca Raton, FL.

Cao, L., & Ramesh, B. (2008). Agile requirements engineering practices: An empirical study. *Software*, *25*(1), 60–67.

Cleland-Huang, J., & Laurent, P. (2014). Requirements in a global world. *IEEE Software*, *31*(6), 34–37.

Dalpiaz, F., & Niu, N. (2020). Requirements engineering in the days of artificial intelligence. *IEEE Software*, *37*(4), 7–10.

Dardenne, A., Van Lamsweerde, A., & Fickas, S. (1993). Goal-directed requirements acquisition. *Science of computer programming*, *20*(1–2), 3–50.

Dassault Systems (2016). CATIA Reqtify. Retrieved from: www.3ds.com/products-services/catia/products/reqtify.

Eberlein, A., & Jiang, L. (2011). Selecting requirements engineering techniques. In P. Laplante (Ed.), *Encyclopedia of Software Engineering*, Published online, pp. 962–978. Boca Raton, FL: Taylor & Francis.

Femmer, H., Fernández, D.M., Juergens, E., Klose, M., Zimmer, I., & Zimmer, J. (2014, June). Rapid requirements checks with requirements smells: Two case studies. In *Proceedings of the 1st International Workshop on Rapid Continuous Software Engineering*, pp. 10–19, ACM.

Forster, T.E. (2003). *Logic, Induction and Sets*. Cambridge University Press, Cambridge.

Gervasi, V., & Nuseibeh, B. (2002). Lightweight validation of natural language requirements. *Software: Practice and Experience*, *32*(2), 113–133.

Gilb, T. (2005). *Competitive Engineering: A Handbook for Systems Engineering, Requirements Engineering, and Software Engineering Using Planguage*. Butterworth-Heinemann. Oxford, UK.

Gotel, O.C.Z. (1995). Contribution structures for requirements traceability. (Doctoral dissertation, University of London).

Howe, J. (2006). The rise of crowdsourcing. *Wired magazine*, *14*(6), 1–4.

Hull, E., Jackson, K., & Dick, J. (2011). Requirements engineering in the problem domain. Eds., E. Hull, K. Jackson, and J. Dick. In *Requirements Engineering*, pp. 93–114. Springer-Verlag, London.

Inayat, I., Salim, S.S., Marczak, S., Daneva, M., & Shamshirband, S. (2015). A systematic literature review on agile requirements engineering practices and challenges. *Computers in human behavior*, *51*, 915–929.

Jeannet, B., & Gaucher, F. (2015). Debugging real-time systems requirements: simulate the "what" before the "how". In Embedded World Conference, Nürnberg, Germany.

Kassab, M., & Laplante, M. (2022). The current and evolving landscape of Requirements Engineering in Practice. *IEEE Software*. Submitted.

Lami, G. (2005). *QuARS: A Tool for Analyzing Requirement (CMU/SEI-2005-TR-014)*. Software Engineering Institute, Carnegie Mellon University. http://resources.sei.cmu.edu/library/asset-view.cfm?AssetID=7681 (accessed 7 January 2017).

Lim, S.L., & Finkelstein, A. (2011). StakeRare: using social networks and collaborative filtering for large-scale requirements elicitation. *IEEE transactions on software engineering*, *38*(3), 707–735.

Laplante, P.A. (2010). Stakeholder analysis for smart grid systems. *In IEEE Reliability Society Annual Technical Report*. http://rs.ieee.org/images/files/Publications/2010/2010-02.pdf (accessed June 2022)

Laplante, P.A., & Kassab, M. (2022). *Requirements engineering for software and systems*. 4th edition. Auerbach Publications.

Li, T., Zhang, F., & Wang, D. (2018). Automatic user preferences elicitation: A data-driven approach. In *International Working Conference on Requirements Engineering: Foundation for Software Quality* (pp. 324–331). Springer, Cham.

Orr, K. (2004). Agile requirements: opportunity or oxymoron? *IEEE Software*, *21*(3), 71–73.

Paetsch, F., Eberlein, A., & Maurer, F. (2003, June). Requirements engineering and agile software development. In *WET ICE 2003. Proceedings. Twelfth IEEE International Workshops on Enabling Technologies: Infrastructure for Collaborative Enterprises, 2003*. (pp. 308–313). IEEE.

Ramesh, B., Cao, L., & Baskerville, R. (2010). Agile requirements engineering practices and challenges: an empirical study. *Information Systems Journal*, *20*(5), 449–480.

Renzel, D., & Klamma, R. (2014). Requirements bazaar: Open-source large scale social requirements engineering in the long tail. *IEEE Computer Society Special Technical Community on Social Networking E-Letter*, *2*.

Savolainen, J., Kuusela, J., & Vilavaara, A. (2010, September). Transition to agile development-rediscovery of important requirements engineering practices. *In 2010 18th IEEE International Requirements Engineering Conference* (pp. 289–294). IEEE.

Scott, W., & Cook, S.C. (2004). A context-free requirements grammar to facilitate automatic assessment. (Doctoral dissertation, UniSA).

Sharma, S., & Pandey, S.K. (2013). Revisiting requirements elicitation techniques. *International Journal of Computer Applications*, *75*(12).

Sharma, S., & Pandey, S.K. (2014). Integrating AI techniques in SDLC: requirements phase perspective. *International Journal of Computer Applications*, *5*(3), 1362–136.

Snijders, R., Dalpiaz, F., Brinkkemper, S., Hosseini, M., Ali, R., & Ozum, A. (2015, August). Refine: A gamified platform for participatory requirements engineering. In *2015 IEEE 1st International Workshop on Crowd-Based Requirements Engineering (CrowdRE)* (pp. 1–6). IEEE.

Spoletini, P., & Ferrari, A. (2017). Requirements elicitation: a look at the future through the lenses of the past. In *2017 IEEE 25th International Requirements Engineering Conference (RE)* (pp. 476–477). IEEE.

Wagner, S., Méndez-Fernández, D., Kalinowski, M., & Felderer, M. (2018). Agile requirements engineering in practice: Status quo and critical problems. *CLEI Electronic Journal*, *21*(1), 15.

Wilson, W.M., Rosenberg, L.H., & Hyatt, L.E. (1997). Automated analysis of requirement specifications. In *Proceedings of the 19th International Conference on Software Engineering*, pp. 161–171.

Whittle, J., Sawyer, P., Bencomo, N., Cheng, B.H., & Bruel, J.M. (2009, August). Relax: Incorporating uncertainty into the specification of self-adaptive systems. In *2009 17th IEEE International Requirements Engineering Conference* (pp. 79–88). IEEE.

Zave, P. (1997). Classification of research efforts in requirements engineering. *ACM Computing Surveys, 29*(4): 315–321.

4 Software Architecture

4.1 INTRODUCTION

The size and complexity of software in systems is on the rise. For example, self-driving cars will require an estimated one billion lines of computer code, which is almost 1,000 times more than the 145,000 lines required by NASA to land Apollo 11 on the moon in 1969. The large proportion of software in systems increases design and operational complexity making them higher risk. To combat this growth in complexity, proper software/system architecture principles need to be used.

Software architecture is significant in enabling software success for many reasons. In addition to helping reduce complexity, the study of software architecture is important to software engineers because it enables them to help stakeholders describe high-level structure (or structures) of the system. It is this structure that facilitates the satisfaction of systemic qualities (such as performance, security, availability, modifiability, etc.). A well-conceived architecture enhances the traceability between the requirements and the technical solutions, which reduces risks associated with building the technical solution. It also serves as a vehicle for communication to the stakeholders of the system under consideration and can serve as a basis for large-scale reuse. Defining software architecture involves a series of decisions based on many factors in a wide range of software development. These decisions are analogous to the load-bearing walls of a building. Once installed, altering them is extremely difficult and expensive. Therefore, each of these decisions can have a considerable impact on satisfying a project's requirements under given constraints.

The fundamental objective of this chapter is to help you develop a clear understanding of the significant role architecture plays in the development of complex systems. Another goal is to get you started in understanding the process for identifying an appropriate architecture before the software is designed, or to be used in redesigning a legacy system. While further study in this subject will be needed to truly become a software architect, we hope to interest you in that study. As with the

DOI: 10.1201/9781003218647-5

foundation and framing of a building, a well designed and developed software architecture can result in a much more reliable, maintainable, and longer lasting system.

4.2 BASIC CONCEPTS

4.2.1 WHAT IS SOFTWARE ARCHITECTURE?

The word architecture is used in many different contexts. We hear terms such as enterprise architecture, cloud computing architecture, IoT system architecture, etc. The common underlying theme across these terms is that we are dealing with structures comprising elements and their properties, and that interact with each other to achieve some useful purpose.

Interactions depend on how these elements are related to each other and the assumptions each makes about what the other provides. For example, in the case of enterprise architecture, we are dealing with an organizational structure. The business processes govern how the various divisions within an organization interact with each other. An interacting division makes assumptions about the services provided by the one it interacts with. A division, however, is less interested in the internal structure and functioning of another division it interacts with. All the divisions are, however, interested in how they communicate, coordinate, and collaborate to fulfill the mission of the enterprise.

The Software Engineering Institute developed recommends the following definition to the "software architecture" term (Bass, Clements, & Kazman, 2021):

> The software architecture of a system is the set of structures needed to reason about the system, which comprise software elements, relations among them, and properties of both.

A representation of (a) structure(s) from a particular perspective is called a view of the system. A view is a partial representation of the system from a given perspective. Software architecture design documentation is comprised of a collection of views. We will visit the discussion on documentation later.

4.2.2 SO, SOFTWARE SYSTEM ARCHITECTURE IS A SET OF STRUCTURES, AND A STRUCTURE CAN BE REPRESENTED THROUGH DIFFERENT VIEWS. WHAT ARE THE TYPES OF VIEWS?

There are three different types of views to represent a structure:

- Static views (e.g., module): These will provide insight into how a system is structured as a set of code units or modules.
- Dynamic views (e.g., Component and Connector (C&C)): These will provide insight into how a system is structured as a set of elements that have runtime behavior (components) and interactions (connectors).
- Physical views (e.g., Allocation): These will show how the system relates to non-software structures in its environment.

Each type of view is suitable for realizing certain characteristics (e.g., quality attributes) of the system. For example, the module view is suitable for discussing "how easy or hard is it to modify the system (modifiability)"; while C&C structures are useful for studying performance, security, and related issues. It is hard to find only one structure that can cover all qualities of the system. Therefore, we need different types of structures when dealing with an architecture of a system. Keep this in mind.

We can draw an analogy between software architecture and the human body. A human body is made of different structures (e.g., skeletal structure, nerves system, digestive system,). In order to be able to describe a person's health you may need to look at all these different structures as no one structure can tell you everything. Now, if you need a representation for a skeletal structure, you may need to have an X-Ray. The X-Ray is a view of the structure in this case. To look at the cardiovascular system you may need a different type of scan. But each structure can be presented via different views.

4.2.3   CAN YOU ILLUSTRATE THE POINT WITH AN EXAMPLE?

Consider a simple program that takes a string as input and produces as an output the same string but with an alternating case of characters ("sofTWareArchitecture" => "SoFtWaReArChItEcTuRe"). One way to present a program's structure is as a set of code units as in Figure 4.1. This structure is a static representation depicting how the code will be distributed across modules while the system is being developed. This structure is suitable to discuss certain software qualities, for example, "modifiability".

On the other hand, one may present how the program is structured as a set of run-time processes (Figure 4.2). This structure is suitable to argue about qualities such as "performance" and "security", for example.

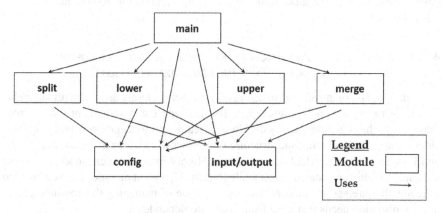

FIGURE 4.1   Static representation for alternating cases program

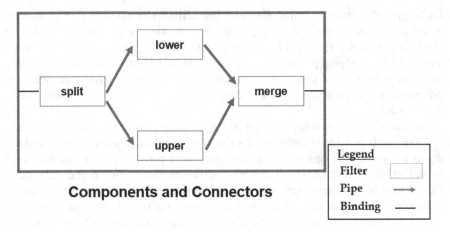

FIGURE 4.2   Dynamic representation for alternating cases program

4.2.4   WHY IS SOFTWARE ARCHITECTURE SIGNIFICANT?

Consider two functionally equivalent implementations of a given system. If you had to choose one and functionality was the only consideration, then any implementation would do. Stated another way, functionality can be satisfied in either of the two system structures. But you may prefer one over the other because it may perform better, is easier to use, is more reliable, and so forth.

So, one of the two structures is more appropriate if qualities such as performance, usability, reliability, etc., are more important to a given system. Therefore, the choice of a structure or an architecture of a system is significantly influenced by the qualities desired from the system.

In addition, architecture serves as a vehicle for communication for the stakeholders of the system under consideration. It is a means used for allocating resources, guiding development work, communicating progress, and negotiating requirements among other things.

4.2.5   CAN SOFTWARE ARCHITECTURE INFLUENCE THE STRUCTURE OF THE DEVELOPMENT TEAM OR EVEN THE ORGANIZATION ITSELF?

Absolutely. The organizational structure of a project is closely aligned with the architecture of the system being developed. To distribute the development work across multiple (perhaps geographically distributed) teams, project managers typically consider a static development-time structure that shows the various modules of a system and how they are dependent on each other. Highly interdependent modules can be allocated to fairly co-located teams while loosely dependent modules can be allocated to teams that are several time zones apart for ease of managing the communication and coordination needs that arise from such interdependence.

Imagine that, after the work allocation has occurred, changes are made to the architecture. If new dependencies among modules emerge that require close collaboration among teams fairly remote from each other, then managing the communication and coordination needs of these teams may become quite challenging. Worst yet, if the modular structure itself changes, then it may require significant changes to how the work was previously allocated to the different teams. This may in turn affect procedures, processes, tools, etc., in place for facilitating the ongoing work.

4.2.6 WHAT MAKES ONE ARCHITECTURE SUPERIOR TO THE OTHER?

The answer depends on the context. If system functionality is the only consideration— and NFRs such as performance, reliability, security, etc., are not, for example— the clear choice may be the architecture that is cheapest to build. If, however, the consideration is an architecture that supports the creation of a product line, those NFRs (such as variability and extensibility) become critical, and neither of the two architectures may be suitable. The suitability of architecture is measured in terms of its fitness to purpose. It is not surprising then that, when the system's fitness to purpose is related to its systemic non-functional requirements, architectures designed to maximize functional cohesion can fail.

4.2.7 WHAT IS THE DIFFERENCE BETWEEN ARCHITECTURE AND PROGRAMMING?

The study and analysis of architecture requires a different approach to thinking. Table 4.1 summarizes the key differences.

4.2.8 WHAT IS THE COURSE OF EVOLUTION OF SOFTWARE ARCHITECTURE AS A DISCIPLINE?

We can answer the question by looking at the key highlights that contributed to the discipline during the 1980s, 1990s, 2000s, and 2010s:

During the 1980s:

- Informal use of box-and-line diagrams to represent the architecture
- Ad hoc application of architectural expertise
- Diverse, uncodified use of architectural patterns and styles
- No identified "architect" role on most projects.

TABLE 4.1
Key Differences between Software Architecture and Software Programming

Architecture focuses on...	Programming focuses on...
interactions among parts	implementations of parts
structural properties	computational properties
system-level performance	algorithmic performance
Outside module boundary	Inside module boundary

During the 1990s:

- Recognition of the value of *architects* in software development organizations
- *Processes* requiring architectural design reviews & explicit architectural documentation
- Use of *product lines*, commercial architectural *standards*, component *integration frameworks*
- *Codification* of vocabulary, notations & tools for architectural design.
- *Books/courses* on software architecture started to emerge.

During the 2000s:

- Incorporation of architectural notions into mainstream design languages (e.g., UML-2), and tools (e.g., Rational Software Architect)
- Methods based on architectural design and refinement (e.g., Model-Driven Architecture)
- Some architecture analysis tools
- Architectural standards for Enterprise Systems (e.g., RM-ODP, TOGAF).

During the 2010s:

- The emergence of Internet-enabled architectures
  - Web services, service-oriented architectures, clouds
- The emergence of supporting platforms, frameworks, and ecosystems
  - .NET, iPhone, Android
- Development in cyber-physical systems
  - Smart grid, energy-aware buildings, collision avoidance systems
- Integration of architecture with software development processes
  - ACDM process, ADD process, Agile+ Architecture.

4.3 ARCHITECTURAL DRIVERS

4.3.1 WHAT ARE ARCHITECTURAL DRIVERS?

Architectural drivers are requirements that shape the software architecture of a project. Architectural drivers include:

- Quality Attributes: Characteristics that the system must possess in addition to the functionality. These are sometimes called non-functional requirements or systemic properties. Of the architectural drivers, quality attributes are the most difficult to uncover, write down, and verify in the design. We have already discussed these in Chapters 2 and 3.
- Constraints: These are conditions that are invariant, and we need to live with them or they are prohibitively expensive to change. They limit the choices available to us. There are different kinds of constraints that can impact the design, for example:

- Business constraints: Conditions that exist within the business that can't be changed and impact design
- Technical constraints: Pre-made design decisions that we have to live with and limit our technical choices.

Quality attributes and constraints drive architectural structure more than functionality:

- Functional requirements describe what the system must do: The architectural design must realize the functional requirements. It is important to understand that the functional requirements by themselves don't inform architectural decisions. In other words, any architecture can support a given functional requirement equally well. But while the architecture is being constructed, then the functional requirements must be realized.

4.3.2   How Do We Generate the Architectural Drivers?

First, the system needs to be aligned with the business context. Then, the overall business context needs to be recognized. Next, the aspects of the business context that impact the system need to be identified and to be articulated in an actionable way. Finally, the relationship between architectural decisions and these business concerns needs to be understood. If the business concerns cannot be fully accommodated, the business context needs to be refined in a way that can be supported by the architecture.

4.3.3   Can You Illustrate This with an Example?

Consider a company that primarily sells sensors and actuators for Home Automation Management Systems (HAMS). The company has software applications that manage a network of these devices, but these applications constitute a "loss leader", i.e., they lose money on the software, but it helps the sale of the hardware devices. Now, the company realizes that the hardware is being commoditized and, over time, the profit margins on the sale of their hardware devices are going to shrink. To sustain their business in the long term, the company decides to create a new HAMS that will be profitable. They wish to accomplish this by doing two things:

1. Reduce internal development costs
2. Expand the market.

The company wishes to build a single HAMS that will manage the different sensors and actuators. With that, the company's internal development costs can be reduced by replacing several of the existing applications with a single one, while the market expansion can be achieved by entering new and emerging geographic markets and opening new sales channels in the form of supporting not only native sensors/actuators but also those from different manufacturers.

The first step is to clearly outline the business goals of the HAMS and their refinements; Table 4.2 restates the business goals and refines these into their corresponding engineering objectives.

TABLE 4.2
Business Goals and Engineering Objectives for HAMS

Business Goal	Engineering Objectives
1. Reduce internal development costs	1. Integrate the four existing management solutions into a single unified HAMS
2. Expand by entering new and emerging geographic markets	2. Support international languages
	3. Comply with regulations impacting life-critical systems, such as fire alarms, to operate within specific latency constraints
3. Open new sales channels in the form of value-added resellers (VARs/VAPs)	4. Support hardware devices from different manufacturers
	5. Support conversions of nonstandard units used by the different hardware devices

From the engineering objectives, we can start defining the functions a product must support. For instance, integration (engineering objective 1) implies the features of existing applications to be integrated (e.g., lighting, HVAC, security, and so forth) must be supported in the new system. This may require innovative ways of displaying information in the user interface and providing fine-grained access control on who is allowed to interact with what part of the system. Supporting international languages (engineering objective 2) implies personalization capabilities. Regulatory policies for safety-critical parts of the system (engineering objective 3) would require alarm handling capabilities for situations that could cause loss of life. Supporting hardware devices from different manufacturers (engineering objective 4) would require dynamic configuration capabilities.

Figure 4.3 presents the functionalities of the system through a use case diagram. In this diagram, we can see that the IoT engineer intends to manage sensor/actuator systems and dynamically reconfigure them. The facilities manager intends to manage alarms generated by sensor/actuator systems that monitor a building. Alarms related to events that could cause loss of life also result in notifications to the public safety system. The system administrator intends to manage the users of the HAMS.

The use cases can be further refined into a set of functional responsibilities that must be fulfilled by the system. The following responsibilities are derived from the use cases shown in Figure 4.3:

1. The system shall send commands to a sensor/actuator device.
2. The system shall receive events from a sensor/actuator device.
3. The system shall perform semantic translation for sensor/actuator device data.
4. The system shall route data to a sensor/actuator device.
5. The system shall evaluate and execute automation rules.
6. The system shall send automation commands.
7. The system shall generate alarm notifications.
8. The system shall display sensor data.

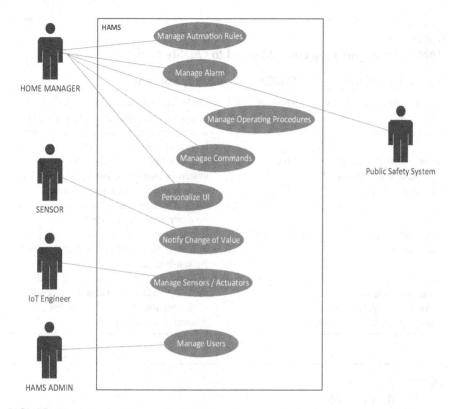

FIGURE 4.3 Use case diagram for HAMS system

9. The system shall capture/relay user commands.
10. The system shall display alarm notifications.
11. The system shall edit/create automation rules.
12. The system shall retrieve data from sensor devices.
13. The system shall store sensor/actuator device configuration.
14. The system shall propagate change of value notifications.
15. The system shall authenticate and authorize users.
16. The system shall persist with automation rules, user preferences, and alarms.

From the engineering objectives, we also can conclude the corresponding quality attributes the end system must exhibit. For example, to support a multitude of IoT hardware devices (engineering objective 4) and consider different languages (engineering objective 2), the system must be modifiable. To support different regulations (engineering objective 3) in different geographic markets, the system must respond to life-threatening events promptly (a performance requirement). It is, therefore, critical that the business goals and their implied quality concerns be fully understood.

Following the Software Engineering Institute's Quality Attribute Workshop, we can elicit concrete scenarios for the quality attributes corresponding to the

TABLE 4.3
HAMS Engineering Objectives Mapped to Quality Attributes

Engineering Objective	Quality	Quality Scenario
Support hardware devices from many different manufacturers	Modifiability	An IoT engineer can integrate a new sensor/actuator device into the system at runtime and the system continues to operate with no downtime or side effects.
Support conversions of nonstandard units used by the different devices	Modifiability	A system administrator configures the system at runtime to handle the units from a newly plugged-in sensor/actuator device and the system continues to operate with no downtime or side effects.
Support international languages	Modifiability	A developer can package a version of the system with new language support in 40 person-hours.
Comply with regulations requiring life-critical systems to operate within specific latency constraints	Performance	A life-critical alarm should be reported to the concerned users within 60 seconds of the occurrence of the event that generated the alarm.

TABLE 4.4
Constraints for HAMS

Category	Factor	Description
Organization	New market segments	Limited experience with some market segments the organization would like to enter.
Technology	Scalability and responsiveness	The system must be scalable to handle a large number of sensor/actuator devices and improve responsiveness.
Product	Performance and scalability	The system must handle a wide range of configurations, say, from 100 sensor/actuator devices to 500,000 sensor/actuator devices.

engineering objectives. Table 4.3 shows a mapping of the engineering objectives to quality attribute scenarios for the HAMS.

Table 4.4 enumerates constraints concluded for the HAMS:

From the functions, quality attribute scenarios, and constraints enumerated in the preceding sections, we distill a list of significant architectural drivers. A prioritized list of such drivers for the HAMS is shown in Table 4.5.

Architectural drivers 1 through 5 relate to the quality attribute scenarios enumerated in Table 4.3. Besides, architectural drivers 1 and 3 also correspond to the dynamic reconfiguration functions, architectural driver 2 corresponds to the internationalization and localization functions, architectural driver 4 corresponds to event

TABLE 4.5
Architectural Drivers

	Architectural Driver	Priority
1	Support for adding new sensor/actuator device	(H, H)
2	International language support	(H, M)
3	Nonstandard unit support	(H, M)
4	Latency of event propagation	(H, H)
5	Latency of alarm propagation	(H, H)
6	Load conditions	(H, H)

management functions, and architectural driver 5 to alarm management functions. Most architectural drivers relate to the factors identified in Table 4.4. For instance, the organizational factor concerning new market segments is reflected in architectural drivers 1 through 5. These drivers consider the flexibility needed to accommodate new sensor/actuator devices and their calibration, language and cultural aspects, and regulatory concerns regarding the responsiveness of the system to safety-critical events. The technological factor related to scalability and responsiveness and the product factor related to performance and scalability is addressed through architectural drivers 4–6.

4.4 ARCHITECTURAL DECISIONS

During the architectural design process, there are certain decisions that need to be made. There are two types of architectural decisions: one involving tactics and the other patterns. In this section, we will discuss each of these two decisions.

4.4.1 WHAT IS A TACTIC?

As we discussed earlier, quality is the totality of characteristics of an entity that bear on its ability to satisfy stated and implied needs. Software quality is an essential and distinguishing attribute of the final architecture product. Tactics on the other hand are measures taken to improve the quality attributes. For example, introducing concurrency for better resource management is a tactic to improve the system's performance. Authorization, authentication, and limited access are tactics to improve security. Modular design, low coupling, and high cohesion are tactics to improve modifiability. Bass et al. (2021) discuss the concept of tactics further and provide a comprehensive list of common tactics for the qualities: Availability, Modifiability, Performance, Security, Testability, and Usability.

4.4.2 WHAT IS A PATTERN?

When designing software architectures, an architect can rely on a set of idiomatic patterns commonly named architectural styles or patterns. A Software Architectural

Pattern defines a family of systems in terms of a pattern of structural organization and behavior. More specifically, an architectural style or pattern determines the vocabulary of components and connectors that can be used in instances of that style, together with a set of constraints on how they can be combined (Aad, 2010). Common architecture patterns include:

- Layered
- Broker
- Model-View-Controller (MVC)
- Pipe and Filter
- Client-Server
- Peer-to-Peer
- Service-Oriented Architecture (SOA)
- Publish-Subscribe
- Shared Data
- Map-Reduce
- Multi-Tier.

If you are interested in learning more about each of the above patterns, Bass et al. (2021) discuss these in more detail.

4.4.3 What Is the State of Practice on Using Architectural Patterns?

Kassab, Mazzara, Lee and Zucchi (2018) provided the results from a comprehensive survey of software professionals to discover the state of practice of using architectural patterns. Some of the notable findings from that survey include:

- Architectural patterns are widely used in software projects. The majority of the software professionals also thought that without patterns it would have not been possible to complete the projects they were involved in. Nevertheless, most of the patterns were applied with changes (by incorporating tactics). Both patterns and tactics are used hand-in-hand when constructing an architectural structure to achieve a comprehensive satisfaction of the architectural drivers.
- The most significant difficulty in adopting patterns in practice is the continuous changes of user requirements or the environment. This finding suggests a better harmonization of the requirements management activities and software architecture process.
- Model-View-Controller was the most commonly used pattern. The most difficult pattern to implement and the most expensive to adopt was peer-to-peer, while the easiest was the client-server.

4.4.4 What Is the Relationship between Tactics and Patterns?

While architectural patterns embody high-level design decisions, an architectural tactic is a design strategy that addresses a particular quality attribute. Tactics serve as the meeting point between the quality attributes and the software architecture. "The structure and behavior of tactics is more local and low-level than the architectural pattern, and therefore must fit into the larger structure and behavior of patterns applied to the same system" (Harrison, Avgeriou, & Zdun, 2010). Implementing a tactic into a pattern may affect the pattern by modifying some of its components, adding some components and connectors, or replicating components and connectors.

Tactics are considered as the building blocks from which architectural patterns are composed (Bass et al., 2021). While an architectural pattern is commonly defined by its components, their interactions and interrelationships, and their semantics, its implementation includes a combination of tactics depending on the pattern's objectives. For example, the Broker pattern implements the modifiability tactic "Use an Intermediary" by introducing the Broker component which acts as an intermediary between clients and servers to provide location-transparent service invocations. Nevertheless, patterns can impact individual tactics by making it easier or more difficult to implement them. Indeed, the changes due to a tactic's implementation within a pattern may, at worst, break the pattern's structure and/or behavior.

In Kassab, El-Boussaidi, & Mili (2012) and Kassab and El-Boussaidi (2013), the authors study quantifying the relationship between qualities, tactics, and patterns.

4.5   ARCHITECTURE DESIGN PROCESS

Several architecture design methods exist. The purpose of any architecture design method is to maintain intellectual control over the design of software systems that:

- Require involvement of and negotiation among multiple stakeholders
- Are often developed by large, distributed teams over extended periods
- Must address multiple possibly conflicting goals and concerns
- Must be maintained for a long period of time.

We will discuss one particular design method developed at the Software Engineering Institute (SEI): Its name is the Attribute-Driven Design (ADD).

4.5.1   WHAT IS ATTRIBUTE-DRIVEN DESIGN?

In ADD, architectural design follows a recursive decomposition process where, at each stage in the decomposition, architectural patterns and tactics are chosen to satisfy a set of architectural drivers. The architecture designed represents the high-level design choices that are documented using multiple structures or views. The nature of the project determines the views.

Architects use the following steps when designing an architecture using the ADD method:

1. Choose the module to decompose. On a greenfield (new) project, the module to start with is usually the whole system. All required inputs (functional requirements, quality requirements, and constraints) for this module should be available.

2. Refine the modules according to these steps:
   a. Choose a subset of architectural drivers from the whole set of drivers that is important for this decomposition.
   b. Choose tactics that satisfy these chosen architectural drivers. Decompose the chosen module into child modules using the chosen tactics.
   c. Allocate functionality to the child modules.
   d. Define interfaces of the child modules.
   e. Prepare the child modules for further decomposition if necessary.
3. Repeat the steps above for every module that needs further decomposition.

4.5.2   Can You Give an Illustrative Example?

Let's revisit the HAMS system. We'll start with the list of architectural drivers that we derived from the business goals. We then begin the process with one of our highest priority architectural drivers (# 1 support for adding a new sensor/actuator device). This driver is related to modifiability quality, and we need to apply modifiability tactics to limit the impact of change and minimize the number of dependencies on the part of the system responsible for integrating new hardware devices. There are three design concerns related to modifiability:

- Localize changes: this relates to adding a new sensor/actuator device.
- Prevention of ripple effects: this relates to minimizing the number of modules affected as a result of adding a new sensor/actuator device.
- Defer binding time: this relates to the time when a new sensor/actuator device is deployed and the ability of non-programmers to manage such deployment.

We address these concerns by creating IoT adaptors for sensor/actuator devices, "an anticipation of expected changes" tactic. We use two additional architectural tactics to minimize the propagation of change. First, we specify a standard interface to be exposed by all IoT adaptors to "maintain existing interfaces". Second, we use the IoT adaptor as an "intermediary" responsible for the semantic translation into a standard format, of all the data received from different sensor/actuator devices. As a side effect, this also addresses architectural driver # 3.

The adaptors are assigned the following responsibilities:

- Send commands to the sensor/actuator device
- Receive events from sensor/actuator device
- Perform semantic translation for sensor/actuator device data.

Instantiating and allocating responsibilities to the adaptors leads to a realization that the HAMS server is still sensitive to a change in the number of sensor/actuator devices it is connected to and must include logic to route commands and data, to and from the correct adaptor. To address this concern we use the "hiding information" tactic introducing an IoT adaptor manager to hide information about the number and type of sensor/actuator devices connected. The IoT adaptor manager together with the adaptors creates a virtual device; that is, for all other components of the HAMS, there is practically one sensor/actuator device to interact with at all times.

Additionally, the IoT adaptor manager uses the following two architectural tactics to address the defer binding time design concern:

- "Runtime registration": This will support plug-and-play operation allowing non-programmers to deploy new sensor/actuator systems.
- "Configuration files": This tactic enables the setting of configuration parameters (such as initial property values) for the sensor/actuator systems at startup.

The IoT adaptor manager is assigned the following responsibilities:

- Configure a sensor/actuator device
- Route data to a sensor/actuator device.

At this stage, architecture driver # 1 is satisfied (and as a side effect so is architecture driver # 3), we next consider architectural drivers 4, 5, and 6 related to the performance quality attribute of the HAMS. There are two design concerns related to these drivers:

- Resource Demand: The arrival of change of property value events from the various sensor/actuator devices and the evaluation of automation rules in response to these events is a source of resource demand.
- Resource Management: The demand on resources may have to be managed to reduce the latency of event and alarm propagation.

To address these concerns, we move the responsibility of rule evaluation and execution, and alarm generation, respectively, to a separate Logic & Reaction (L&R) component and an alarm component. These components running outside the automation server can now easily be moved to dedicated execution nodes if necessary. In doing so, we are making use of the increased available resources tactic to address the resource management concern and the reduced computational overhead tactic to address the resource demand concern.

We use an additional tactic to address the resource management concern. This tactic relies on introducing "concurrency" to reduce delays in processing time. Concurrency is used inside the L&R and alarm components to perform simultaneous rule evaluations.

We assign the following responsibilities to the L&R and alarm components:

- Evaluate and execute automation rules
- Send automation commands

Besides, we assign the following responsibility to the alarm component:

- Generate alarm notifications

The results of applying these tactics are shown in Figure 4.4.

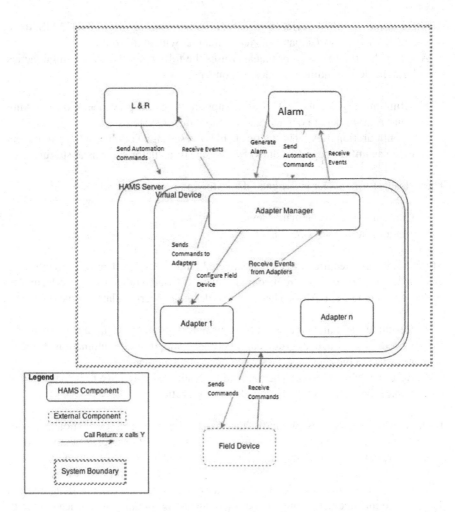

FIGURE 4.4   HAMS architecture after applying architectural drivers 1, 4, 5, and 6

We finally consider architectural driver 2 related to international language support for the HAMS. There are two design concerns for this driver:

- Localize changes: This relates to changing the user interface to deal with a new language and culture.
- Prevention of ripple effects: This relates to minimizing the number of modules affected as a result of changing the user interface.

We address these concerns using the following modifiability tactics to address localize changes and prevention of ripple effects design concerns:

- Anticipation of expected changes: Changes to the user interface (UI) are localized to the presentation component.

- Intermediary: The presentation component acts as an intermediary preventing ripple effects from changes to the UI from propagating to the rest of the application.

We assign the following responsibilities to the presentation component:

- Display device data
- Capture/relay user commands
- Display alarm conditions
- Edit/create automation rules.

The results of applying these tactics are shown in Figure 4.5.

The architecture elaboration process we used here is iterative. Moreover, the tactics we choose to implement can very often have a negative impact on the quality attributes they do not target specifically. In the case of the HAMS, we focused on modifiability and performance tactics which can have a negative impact on each other. We revisited the performance and modifiability drivers next to address these issues.

Introducing performance tactics resulted in the creation of multiple components (Rule Manager and Alarm Manager, for instance) that now depend on the virtual device. Therefore, based on its current structure, we can predict that some changes to the virtual device have the potential to propagate to several other components of the HAMS. We would like to minimize the ripple effect of these changes. To achieve this objective, we introduce a Publish-Subscribe bus. We allocate the following responsibility to the Publish-Subscribe bus:

- Propagate change of value notifications.

By examining the current system structure, it can be seen that every time the Rule Manager and the Alarm Manager needs to query a field device, it needs to make a call that traverses multiple components along the way to the field device.

Since crossing component boundaries typically introduces computational overhead, and because the querying latency of field devices is a constraint over which we have no control, we decompose the virtual device and introduce a cache component to improve device querying performance.

This cache provides field device property values to the system, saving part of the performance cost incurred when querying the actual field devices. The performance gains are seen because we reduce the number of component and machine boundaries traversed for each query. A cache is an application of *maintaining multiple copies of data* performance tactics. We allocate the following functional responsibilities to the cache:

- Retrieve data from a field device
- Store field device configuration.

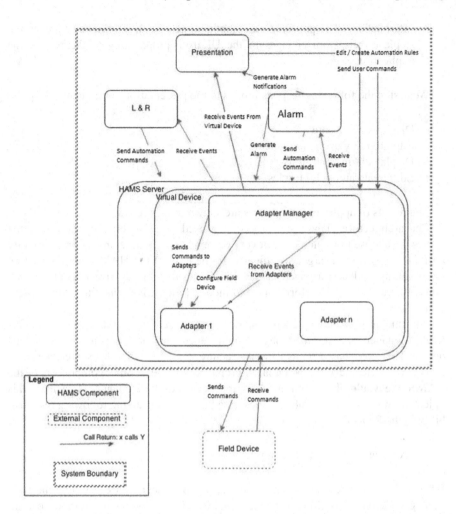

FIGURE 4.5 HAMS architecture after applying all architectural drivers

4.6   ARCHITECTURE DOCUMENTATION

Architecture documentation aims at communicating architectural information and ideas. If you can't explain it to someone, it has little value. Failure to document well is often a symptom of sloppy and incomplete thinking. These principles are especially true with respect to software architecture.

4.6.1   IS THERE A RECOMMENDED TEMPLATE TO FOLLOW TO DOCUMENT THE ARCHITECTURE?

While discussing the HAMS system, we presented its architecture through one view (component-connector view). But in practice, architects rely on more than one view to present the architecture. Which views you document depends on the uses you expect to make of the documentation. For instance, developers are concerned with how a system is structured so they can understand how elements they are developing depend on the rest of the system and how these elements interact with each other at runtime. A module view can be used to depict the static development time structure of the system showing its decomposition into different modules along with their inter-dependencies. A C&C view, on the other hand, can be used to depict the dynamic runtime structure of the system showing its decomposition into elements (such as components, tasks, processes, threads, etc.) that have runtime behavior and inter-action. The module view can be used for reasoning about modifiability concerns and allows developers to understand how changes they make can impact others and how changes others make can impact them. The C&C view can be used for addressing concerns related to many of the runtime qualities of a system such as performance, security, and availability.

The Views and Beyond approach (Clements, Garlan, Little, Nord, & Stafford , 2011) suggests each view be documented with the following items:

- Primary Presentation: A visual representation that shows the elements in a structure and the relationships among them.
- Element Catalog: Details at least those elements and relations depicted in the primary presentation; these details include the interfaces of the elements and how these elements behave at runtime.
- Context Diagram: Shows how the system depicted in the view relates to its environment.
- Architecture Background: Explains the design rationale, analysis results, and assumptions.

For more details on templates that can be used to follow the architecture, the book "Documenting Software Architecture: views and beyond" by Clements et al. (2011) is an excellent reference.

4.6.2   What Are Some Points to Keep in Mind While Documenting Architecture?

- Architectures are not just pretty pictures: They convey design decisions, including structure and semantics.
- Bad documentation usually indicates fuzzy thinking: If you can't write it down, you probably don't understand what you are doing.
- Documentation is not an afterthought: All stages of architecture should have *appropriate* documentation.
- Different kinds of documentation are good for at different times.
- Pictures are not enough; documentation needs rationale, context, and prose to interpret the pictures.
- An architect must document software-intensive systems from at least the three perspectives, we discussed earlier:
  - Static: How is it structured as a set of code units?

- Dynamic: How is it structured as a set of elements that have runtime behavior and interactions?
- Physical: How does it relate to non-software structures in its environment?

4.6.3 WHICH NOTATIONS CAN AN ARCHITECT USE TO REPRESENT THE ARCHITECTURE?

- Informal (e.g., boxes and lines – similar to how we presented the HAMS system): This can be effective if following guidelines for good architectural documentation.
- Semi-formal (e.g., UML): Usually general purpose (not specifically for architecture); so needs to be used with care.
- Formal (e.g., Acme): Usually specialized for architecture (ADLs) and provides analytic leverage, style support, code generation, etc.

4.6.4 WHAT ARE THE BASIC PRINCIPLES OF GOOD ARCHITECTURAL DOCUMENTATION?

- The graphics representing the architecture shall communicate semantics effectively:
  - Elements have different shapes/colors and a key/legend to explain what these mean
  - Connector types are clearly distinguished.
- The diagrams representing the architecture shall contain an appropriate level of detail:
  - Each view of architecture fits on a page
  - Separated into views, where necessary
  - Use of hierarchy.
- There must be a clear distinction between view types:
  - Separation of concerns
  - But with mappings, where appropriate.

4.6.5 WHAT ARE SOME OF THE COMMON BAD PRACTICES WHEN DOCUMENTING SOFTWARE ARCHITECTURE?

- The lines all look the same: arrows don't mean anything or could mean many things.
- There is too little or too much detail: Low-level implementation concerns are inappropriately mixed with architectural abstraction.
- Ambiguous box-and-line diagrams:
  - There is no key or legend
  - Diagrams are missing relationships
  - Inconsistent use of notations.

- Incomplete prose: Poorly described designs
  - Unclear project context
  - Poor justification of rationale
  - No discussion of alternatives.

4.7 ARCHITECTURE EVALUATION

4.7.1 WHY IS IT IMPORTANT TO EVALUATE THE ARCHITECTURE?

Software architecture is the earliest life cycle artifact that embodies significant design decisions. Architectural design can have major consequences in estimation and project planning (construction). The potential time, cost, and consequences of failure justify the cost of design evaluation. The average architecture review can pay back at least 12 times its cost (Bass et al., 2021). Other beneficial side effects to the evaluation process include:

- Cross-organizational learning is enhanced.
- Architectural reviews assist organizational change.
- Greater opportunities exist to find different defects in integration and system tests.

4.7.2 WHAT ARE THE CHALLENGES IN EVALUATING ARCHITECTURE?

Architectural evaluation requires expertise to handle a range of possible scenarios. For example, the requirements for a system may not be documented or known. Or there could be a lack of common understanding of the architectural design because:

- Rationale not documented
- Design not documented or poorly documented
- No standard notion to describe the design

4.7.2.1 What Are the Different Ways to Evaluate the Architecture?
- Simulations: These are usually tool-based and generally aid in understanding dynamic aspects of the system,
- Formal model analysis: Architectures are formally modeled, and the model checked or verified. Generally, focuses on establishing the existence or absence of various properties.
  - Prototype: Build a facsimile and refine the architecture based on the experience.
  - Architecture reviews:
    - Internal
    - Corporate (e.g., "architecture review board")
    - External (e.g., the Architecture tradeoff analysis method (ATAM)).

FURTHER READING

Aad G., Abbott B., Abdallah J., Abdelalim A., Abdesselam A., Abdinov O., Abi B., Abolins M., Abramowicz H., Abreu H. et al (2010). The ATLAS simulation infrastructure. *The European Physical Journal C, 70*(3), 823–874.

Bass, L., Clements, P., & Kazman, R. (2021). *Software architecture in practice*. Addison-Wesley Professional.

Clements, P., Garlan, D., Little, R., Nord, R., & Stafford, J. (2011). Documenting software architectures: views and beyond. In *25th International Conference on Software Engineering, 2003. Proceedings.* (pp. 740–741). IEEE.

Harrison, N.B., Avgeriou, P., & Zdun, U. (2010, April). On the impact of fault tolerance tactics on architecture patterns. In *Proceedings of the 2nd International Workshop on Software Engineering for Resilient Systems* (pp. 12–21).

Kassab, M., & El-Boussaidi, G. (2013, January). Towards Quantifying Quality, Tactics and Architectural Patterns Interactions (S). In *SEKE* (pp. 441–446).

Kassab, M., El-Boussaidi, G., & Mili, H. (2012). A quantitative evaluation of the impact of architectural patterns on quality requirements. In *Software Engineering Research, Management and Applications 2011* (pp. 173–184). Springer, Berlin, Heidelberg.

Kassab, M., Mazzara, M., Lee, J., & Succi, G. (2018). Software architectural patterns in practice: an empirical study. *Innovations in Systems and Software Engineering, 14*(4), 263–271.

5 Designing Software

5.1 INTRODUCTION

Mature engineering disciplines have handbooks that describe successful solutions to known problems. For example, automobile designers don't design cars using the laws of physics; they adapt adequate solutions from the handbook known to work well enough.

The extra few percent of performance available by starting from scratch is typically not worth the cost.

Software engineering has been criticized for not having the same kind of underlying rigor as other engineering disciplines. But while it may be true that there are few formulaic principles, there are fundamental rules that form the basis of sound software engineering practice. These rules can form the basis of the handbook for software engineering.

If software is to become an engineering discipline, successful practices must be systematically documented and widely disseminated. The following sections describe the most general and prevalent of these concepts and practices for software design.

5.2 SOFTWARE DESIGN CONCEPTS

5.2.1 WHAT IS SOFTWARE DESIGN?

Software design involves identifying the components of the software design, their inner workings, and their interfaces from the SRS and architecture. The principal artifact of this activity is the software design specification (SDS), which is also referred to as a software design description (SDD).

5.2.2 WHAT IS THE DIFFERENCE BETWEEN SOFTWARE ARCHITECTURE AND SOFTWARE DESIGN?

Software architecture deals with constructing the high-level structure of the system with its components and modules. While software design is concerned about designing the individual modules and components. Software architecture manages

DOI: 10.1201/9781003218647-6

uncertainty while software design avoids uncertainty. Software architecture is a pre-requisite to the software design, while software design is a prerequisite to the soft-ware implementation.

5.2.3 WHAT ARE THE PRINCIPAL ACTIVITIES OF SOFTWARE DESIGN?

The activities for the software design phase go hand-in-hand with the activities of the software architecture. While the architecture deals with the high-level structure of the system, the activities of the software design deal with individual components/ modules that are identified from the software architecture phase. These activities include:

- Performing hardware/software trade-off analysis
- Designing interfaces to external components (hardware, software, and user interfaces)
- Designing interfaces between components
- Making the determination between centralized or distributed processing schemes
- Determining concurrency of execution
- Designing control strategies
- Determining data storage, maintenance, and allocation strategy
- Designing databases, structures, and handling routines
- Designing the startup and shutdown processing
- Designing algorithms and functional processing
- Designing error processing and error message handling
- Conducting performance analyses
- Specifying the physical location of components and data
- Creating documentation for the system
- Conducting internal reviews
- Developing the detailed design for the components identified in the software architecture
- Documenting the software architecture in the form of the SDS
- Presenting the detail design information at a formal design review.

This is an intimidating set of tasks that is further complicated as many of these must occur in parallel or be iterated several times. Moreover, because clearly more than one individual must be involved in these activities, problems of working on a project team come into play.

There are no rules of thumb per se for conducting these tasks. Instead, it takes many years of practice, experience, learning from the experience of others, and good judgment to guide the software engineer through this maze of tasks.

5.3 BASIC SOFTWARE ENGINEERING PRINCIPLES

5.3.1 HOW DO RIGOR AND FORMALITY ENTER INTO SOFTWARE ENGINEERING?

Because software development is a creative activity, there is an inherent tendency toward informal ad hoc techniques in software specification, design, and coding. But the informal approach is contrary to good software engineering practices.

Rigor in software engineering requires the use of mathematical techniques. Formality is a higher form of rigor in which precise engineering approaches are used. For example, imaging systems require the use of rigorous mathematical specification in the description of image acquisition, filtering, enhancement, etc. But the existence of mathematical equations in the requirements or design does not imply an overall formal software engineering approach. In the case of the baggage inspection system, formality further requires that there be an underlying algorithmic approach to the specification, design, coding, and documentation of the software.

5.3.2 WHAT IS SEPARATION OF CONCERNS?

Separation of concerns is a kind of divide and conquer strategy that software engineers use. There are various ways in which separation of concerns can be achieved. In terms of software design and coding, it is found in modularization of code and in object-oriented design. There may be separation in time; for example, developing a schedule for a collection of periodic computing tasks with different periods.

Yet another way of separating concerns is in dealing with qualities. For example, it may be helpful to address the fault tolerance of a system while ignoring other qualities. However, it must be remembered that many of the qualities of software are interrelated, and it is generally impossible to affect one without affecting the other, possibly adversely.

5.3.3 CAN MODULAR DESIGN LEAD TO SEPARATION OF CONCERNS?

Some separation of concerns can be achieved in software through modular design. Modular design, first proposed by Parnas (1972), involves the decomposition of software behavior in encapsulated software units. Separation of concerns can be achieved in either object-oriented or procedurally oriented programming languages.

Modularity is achieved by grouping together logically related elements, such as statements, procedures, variable declarations, object attributes, and so on, in increasingly greater levels of detail (Figure 5.1).

The main objectives in seeking modularity are to foster high cohesion and low coupling. With respect to the code units, cohesion represents intra-module connectivity and coupling represents inter-module connectivity. Coupling and cohesion can be illustrated informally as in Figure 5.2, which shows software structures with high cohesion and low coupling (Figure 5.2a) and low cohesion and high coupling (Figure 5.2b). The inside squares represent statements or data, and the arcs indicate functional dependency.

5.3.4 WHAT IS COHESION?

Cohesion relates to the relationship of the elements of a module. Constantine and Yourdon identified seven levels of cohesion in order of strength (Pressman & Maxim, 2019):

- Coincidental: Parts of the module are not related but are simply bundled into a single module.

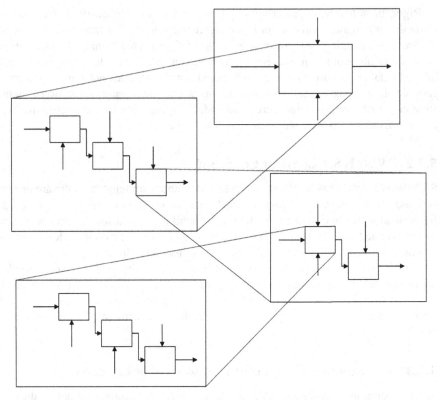

FIGURE 5.1 Modular decomposition of code units. The arrows represent inputs and outputs in the procedural paradigm. In the object-oriented paradigm, they represent method invocations or messages. The boxes represent encapsulated data and procedures in the procedural paradigm. In the object-oriented paradigm they represent classes

- Logical: Parts that perform similar tasks are put together in a module.
- Temporal: Tasks that execute within the same time span are brought together.
- Procedural: The elements of a module make up a single control sequence.
- Communicational: All elements of a module act on the same area of a data structure.
- Sequential: The output of one part of a module serves as input for another part.
- Functional: Each part of the module is necessary for the execution of a single function.

High cohesion implies that each module represents a single part of the problem solution. Therefore, if the system ever needs modification, then the part that needs to be modified exists in a single place, making it easier to change.

5.3.5 WHAT IS COUPLING?

Coupling relates to the relationships between the modules themselves. There is great benefit in reducing coupling so that changes made to one code unit do not propagate

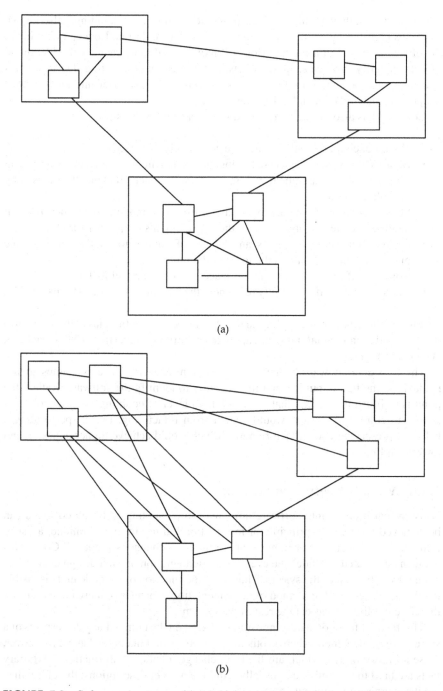

(a)

(b)

FIGURE 5.2 Software structures with (a) high cohesion and low coupling and (b) low cohesion and high coupling. The inside squares represent statements or data, and the arcs indicate functional dependency

to others; that is, they are hidden. This principle of "information hiding," also known as Parnas partitioning, is the cornerstone of all software design. Low coupling limits the effects of errors in a module (lower "ripple effect") and reduces the likelihood of data integrity problems. In some cases, however, high coupling due to control structures may be necessary. For example, in most graphical user interfaces control coupling is unavoidable, and indeed desirable.

Coupling has also been characterized in increasing levels as follows:

- No direct coupling: All modules are completely unrelated.
- Data: When all arguments are homogeneous data items; that is, every argument is either a simple argument or data structure in which all elements are used by the called module.
- Stamp: When a data structure is passed from one module to another, but that module operates on only some of the data elements of the structure.
- Control: One module passes an element of control to another; that is, one module explicitly controls the logic of the other.
- Common: If two modules both have access to the same global data.
- Content: One module directly references the contents of another (Parnas, 1972).

To further illustrate both coupling and cohesion, consider the class structure for a widely used commercial imaging application program interface (API) package, depicted in Figure 5.3.

The class diagram was obtained through design recovery. The class names are not readable in the figure, but it is not the intention to identify the software. Rather, the point is to illustrate the fact that there is a high level of coupling and low cohesion in the structure. This design would benefit from refactoring; that is, performing a behavior preserving code transformation, which would achieve higher cohesion and lower coupling.

5.3.6 WHAT IS PARNAS PARTITIONING?

Software partitioning into software units with low coupling and high cohesion can be achieved through the principle of information hiding. In this technique, a list of difficult design decisions or things that are likely to change is prepared. Code units are then designated to "hide" the eventual implementation of each design decision or feature from the rest of the system. Thus, only the function of the code units is visible to other modules, not the method of implementation. Changes in these code units are therefore not likely to affect the rest of the system.

This form of functional decomposition is based on the notion that some aspects of a system are fundamental, whereas others are arbitrary and likely to change. Moreover, those arbitrary things, which are likely to change, contain "information." Arbitrary facts are hard to remember and usually require lengthier descriptions; therefore, they are the sources of complexity.

Parnas partitioning "hides" the implementation details of software features, design decisions, low-level drivers, etc., in order to limit the scope of impact of

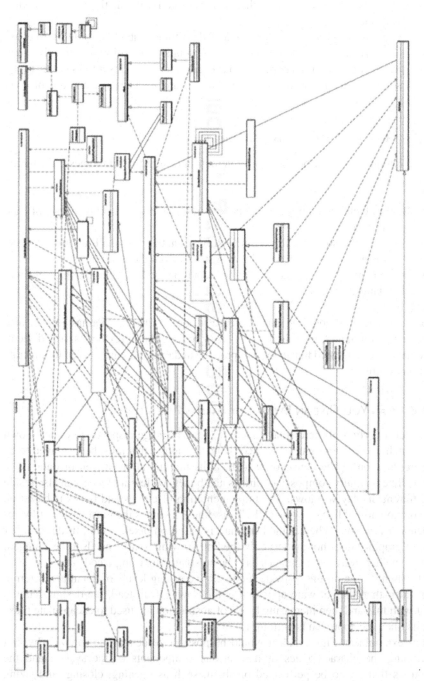

FIGURE 5.3 The class structure for the API of a widely deployed imaging package

future changes or corrections. By partitioning things likely to change, only that module need be touched when a change is required without the need to modify unaffected code.

This technique is particularly applicable and useful in embedded systems. Because they are so directly tied to hardware, it is important to partition and localize each implementation detail with a particular hardware interface. This allows easier future modification due to hardware interface changes and reduces the amount of code affected.

5.3.7 How Do I Do Parnas Partitioning?

The following steps can be used to implement a design that embodies information hiding.

- Begin by characterizing the likely changes. Estimate the probabilities of each type of change.
- Organize the software to confine likely changes to a small amount of code. Provide an "abstract interface" that abstracts from the differences.
- Implement "objects;" that is, abstract data types and modules that hide changeable data structures.

These steps reduce coupling and increase module cohesion. Parnas (1972) also indicated that although module design is easy to describe in textbooks, it is difficult to achieve. He suggested that extensive practice and examples are needed to illustrate the point correctly.

5.3.8 Can You Give an Example of Parnas Partitioning?

Consider a portion of the display function of the baggage inspection system shown in hierarchical form in Figure 5.4. It consists of graphics that must be displayed (e.g., a representation of the conveyor system, units moving along it, sensor data, etc.), which are essentially composed of circles and boxes. Different objects can also reside in different display windows. The implementation of circles and boxes is based on the composition of line drawing calls. Thus, line drawing is the most basic hardware-dependent function. Whether the hardware is based on pixel, vector, turtle, or other type of graphics does not matter; only the line drawing routine needs to be changed. Hence, the hardware dependencies have been isolated to a single code unit.

If, when designing the software modules, increasing levels of detail are deferred until later, then the software approach is top-down. If, instead, the design detail is dealt with first and then increasing levels of abstraction are used to encapsulate those details, then the approach is bottom-up.

For example, in Figure 5.4, it would be possible to design the software by first describing the characteristics of the various components of the system and the functions that are to be performed on them, such as opening, closing, and sizing windows. Then the window functionality could be decomposed into its constituent

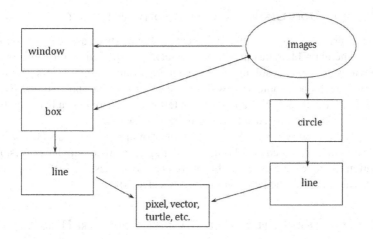

FIGURE 5.4   Parnas partitioning of graphics rendering software

parts such as boxes and text. These could be decomposed still further; for example, all boxes consist of lines and so on. The top-down refinement continues until the lowest level of detail needed for code development has been reached.

Alternatively, it is possible to begin by encapsulating the details of the most volatile part of the system, the hardware implementation of a single line or pixel, into a single code unit. Then working upward, increasing levels of abstraction are created until the system requirements are satisfied. This is a bottom-up approach to design.

5.3.9   CAN YOU DO PARNAS PARTITIONING IN OBJECT-ORIENTED DESIGN?

Yes. In the object-oriented paradigm, Parnas partitioning is a form of a design technique called protected variation, which is one of the "GRASP" principles (to be discussed later in this chapter).

5.3.10   HOW CAN CHANGES BE ANTICIPATED IN SOFTWARE DESIGNS?

It has been mentioned that software products are subject to frequent change either to support new hardware or software requirements or to repair defects. A high maintainability level of the software product is one of the hallmarks of outstanding commercial software.

Engineers know that their systems are frequently subject to changes in hardware, algorithms, and even application. Therefore, these systems must be designed in such a way to facilitate changes without degrading the other desirable properties of the software.

Anticipation of change can be achieved in the software design through appropriate techniques, through the adoption of an appropriate software life cycle model and associated methodologies, and through appropriate management practices.

5.3.11   How Does Generality Apply to Software Design?

In solving a problem, the Principle of Generality can be stated as the intent to look for the more general problem that may be hidden behind a specialization of that problem. In an obvious example, designing the baggage inspection system is less general than designing it to be adaptable to a wide range of inspection applications. Although generalized solutions may be more costly in terms of the problem at hand, in the long run the costs of a generalized solution may be worthwhile.

Generality can be achieved through a number of approaches associated with procedural and object-oriented paradigms. For example, in procedural languages, Parnas information hiding can be used. In object-oriented languages, the Liskov Substitution Principle can be used. The latter will be discussed shortly.

5.3.12   How Does Incrementality Manifest in Software Design?

Incrementality involves a software approach in which progressively larger increments of the desired product are developed. Each increment provides additional functionality, which brings the product closer to the final one.

Each increment also offers an opportunity for demonstration of the product to the customer for the purposes of gathering requirements and refining the look and feel of the product.

5.4   SOFTWARE DESIGN MODELING

5.4.1   What Standard Methodologies Can Be Used for Software Design?

Two methodologies, process- or procedural-oriented and object-oriented design (OOD), are related to structured analysis (SA) and object-oriented analysis (OOA), respectively, and can be used to begin to perform the design activities from the SRS produced by either SA or structured design (SD). Each methodology seeks to arrive at a model containing small, detailed components.

5.4.2   What Is Procedural-Oriented Design?

Procedural-oriented design methodologies, such as SD, involve top-down or bottom-up approaches centered on procedural languages such as C and Fortran. The most common of these approaches utilizes design decomposition via Parnas partitioning.

Object-oriented languages provide a natural environment for information hiding, through encapsulation. The state, or data, and behavior, or methods, of objects are encapsulated and accessed only via a published interface or private methods. For example, in image processing systems, one may wish to define a class of type pixel, with characteristics (attributes) describing its position, color, brightness, and so on, and operations that can be applied to a pixel such as add, activate, and deactivate. The engineer may then wish to define objects of type image as a collection of pixels with other attributes. In some cases, expression of system functionality is easier to do in an object-oriented manner. We will discuss OOD shortly.

5.4.3 What Is SD?

SD is the companion methodology to SA. SD is a systematic approach concerned with the specification of the software architecture and involves a number of techniques, strategies, and tools. SD provides a step-by-step design process that is intended to improve software quality, reduce risk of failure, and increase reliability, flexibility, maintainability, and effectiveness.

5.4.4 What Is the Difference between SA and SD?

SA is related to SD in the same way that a requirements representation is related to the software design; that is, the former is functional and flat, and the latter is modular and hierarchical. Data structure diagrams are then used to give information about logical relationships in complex data structures.

5.4.5 How Do I Go from SA to SD?

The transition mechanisms from SA to SD are manual and involve significant analysis and trade-offs of alternative approaches. Normally, SD proceeds from SA in the following manner. Once the context diagram is drawn, a set of Data Flow Diagrams (DFDs) is developed. The first DFD, the Level 0 diagram, shows the highest level of system abstraction. Decomposing processes to lower and lower levels until they are ready for detailed design renders new DFDs with successive levels of increasing detail. This decomposition process is called *leveling*.

In a typical DFD, boxes represent terminators that are labeled with a noun phrase that describes the system, agent, or device from which data enters or exits. Each process, depicted by a circle, is labeled as a verb phrase describing the operation to be performed on the data although it may be labeled with the name of a system or operation that manipulates the data. Solid arcs are used to connect terminators to processes and between processes to indicate the flow of data through the system. Each arc is labeled with a noun phrase that describes the data. Dashed arcs are discussed later. Parallel lines indicate data stores, which are labeled by a noun phrase naming the file, database, or repository where the system stores the data.

Each DFD should have between 3 and 9 processes only. The descriptions for the lowest level, or primitive, processes are called process specifications, or P-Specs, and are expressed in either structured English, pseudo-code, decision tables, or decision trees and are used to describe the logic and policy of the program (Figure 5.5). Returning to the baggage inspection system example, Figure 5.6 shows the Level 0 DFD. Here the details of the system are given at a high level.

First, the system reacts to the arrival of a new product by confirming that the image data is available. Next, the system captures the image by buffering the raw data from the capture device to a file. Preprocessing of the raw data is performed to produce an image frame to be used for classification and generation of the appropriate control

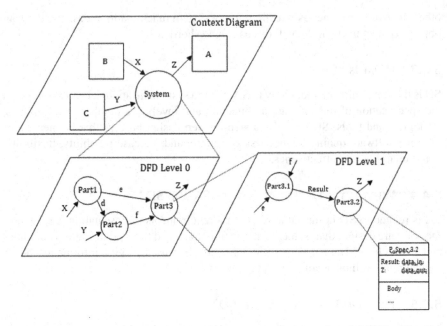

FIGURE 5.5   Context diagram evolution from context diagram to Level 0 DFD to Level 1 DFD and, finally, to a P-Spec, which is suitable for coding

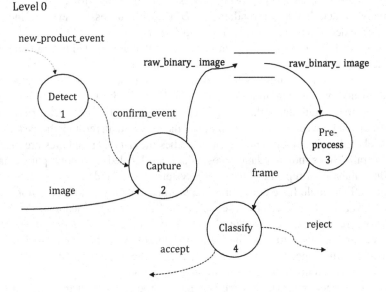

FIGURE 5.6   The Level 0 DFD for the baggage inspection system. The dashed arcs represent control flows, which are described later

signals to the conveyor system. Proceeding to the next level will provide then more detail for Processes 1, 2, 3, and 4.

5.4.6   What Is OOD?

OOD is an approach to systems design that views the system components as objects and data processes, control processes, and data stores that are encapsulated within objects. Early forays into OOD were led by attempts to reuse some of the better features of conventional methodologies, such as the DFDs and entity relationship models by reinterpreting them in the context of object-oriented languages. This approach can be seen in the UML. Over the last two decades, the object-oriented framework has gained significant acceptance in the software engineering community.

5.4.7   What Are the Benefits of Object Orientation?

Previous discussions highlighted some considerations concerning the appropriateness of the object-oriented paradigm to various application areas. The real advantages of applying object-oriented paradigms are the future extensibility and reuse that can be attained, and the relative ease of future changes.

Software systems are subject to near-continuous change: requirements change, merge, emerge, and mutate; target languages, platforms, and architectures change; and, most significantly, employment of the software in practice changes. This flexibility places considerable burden on the software design: how can systems that must support such widespread change be built without compromising quality? There are several basic rules of OOD that help us achieve these benefits.

5.4.8   What Are the Basic Rules of OOD?

The following list of rules is widely accepted. All of these rules, except where noted, are from Martin (2002).

- *The Open/Closed Principle*: Software entities (classes, modules, etc.) should be open for extension, but closed for modification (Meyer, 2000).
- *The Liskov Substitution Principle*: Derived classes must be usable through the base class interface without the need for the user to know the difference (Liskov & Wing, 1994).
- *The Dependency Inversion Principle*: Details should depend upon abstractions. Abstractions should not depend upon details.
- *The Interface Segregation Principle*: Many client-specific interfaces are better than one general purpose interface.
- *The Reuse/Release Equivalency Principle*: The granule of reuse is the same as the granule of release. Only components that are released through a tracking system can be effectively reused.
- *The Common Closure Principle*: Classes that change together belong together.
- *The Common Reuse Principle*: Classes that aren't reused together should not be grouped together.

- *The Acyclic Dependencies Principle*: The dependency structure for released components must be a directed acyclic graph. There can be no cycles.
- *The Stable Dependencies Principle*: Dependencies between released categories must run in the direction of stability. The dependee must be more stable than the depender.
- *The Stable Abstractions Principle*: The more stable a class category is, the more it must consist of abstract classes. A completely stable category should consist of nothing but abstract classes.
- *Once and only once (OAOO)*: Any aspect of a software system—be it an algorithm, a set of constants, documentation, or logic—should exist in only one place (Beck & Andres, 2004).

A detailed discussion of these can be found in the aforementioned references.

5.4.9   WHAT IS THE UML?

The UML is widely accepted as the *de facto* standard language for the specification and design of software-intensive systems using an object-oriented approach. By bringing together the "best-of-breed" in specification techniques, the UML has become a family of languages (diagram types). Users can choose which members of the family are suitable for their domain.

5.4.10   HOW DOES THE UML HELP US WITH SOFTWARE DESIGN?

The UML is a graphical language based upon the premise that any system can be composed of communities of interacting entities and that various aspects of those entities, and their communication, can be described using the set of nine diagrams: use case, sequence, collaboration, statechart, activity, class, object, component, and deployment. Of these, five render behavioral views (use case, sequence, collaboration, statechart, and activity) while the remaining ones are concerned with architectural or static aspects.

With respect to embedded systems, these behavioral models are of interest. The use case diagrams document the dialog between external actors and the system under development. Sequence and collaboration diagrams describe interactions between objects. Activity diagrams illustrate the flow-of-control between objects. Statecharts represent the internal dynamics of active objects. The principal artifacts generated when using the UML and their relationships are shown in Figure 5.7.

While not aimed specifically at embedded system design, some notion of time has been included in the UML through the use of sequence diagrams. Other features of UML include:

- Base classes that provide the foundation for UML modeling constructs
- Object constraint language, a formal method that can be used to better describe object interactions
- Meta-modeling that allows users to model systems from four viewpoints:

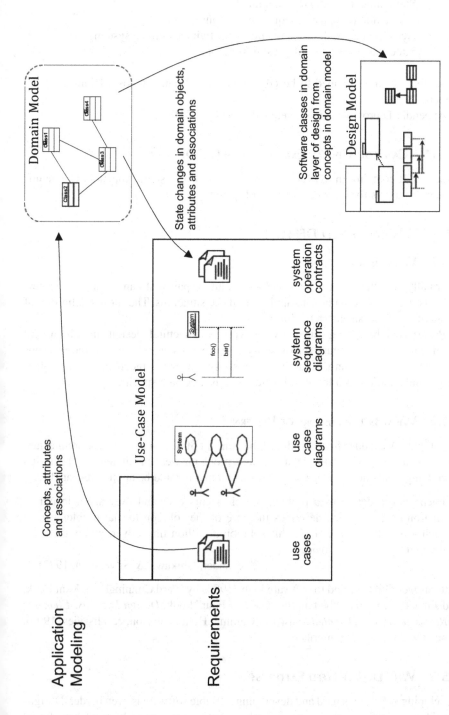

FIGURE 5.7   The UML and its role in specification and design (Adapted from Larman, 2004)

- Static models (e.g., class diagrams)
- Interaction (e.g., using sequence diagrams)
- Activity (i.e., to describe the flow of activities within a system)
- State (i.e., to create FSMs using statecharts).

These features are intended to be complementary (France, Ghosh, Dinh-Trong, & Solberg, 2006).

Appendix D provides a more detailed overview of UML.

5.4.11   CAN YOU GIVE AN EXAMPLE OF AN OOD?

Appendix B contains an SDS for the wet well control system. Appendix C contains the object models for the wet well control system.

5.5   PATTERN-BASED DESIGN

5.5.1   WHAT IS A PATTERN?

Informally, a pattern is a named problem–solution pair that can be applied in new contexts, with advice on how to apply it in those situations. The formal definition of a pattern is not consistent in the literature.

Patterns can be distinguished as three types: architectural, design, and idioms. An architectural pattern occurs across subsystems (we discussed these in the previous chapter), a design pattern occurs within a subsystem but is independent of the language, and an idiom is a low-level pattern that is language-specific.

5.5.2   WHAT IS THE HISTORY OF PATTERNS?

Christopher Alexander first introduced the concept of design patterns for architecture and town planning. He realized that the same problems were encountered in the design of buildings and once an elegant solution was found it could be applied repeatedly.

Each pattern describes a problem which occurs over and over again in our environment, and then describes the core of the solution to that problem, in such a way that you can use this solution a million times over, without ever doing it the same way twice.

(Alexander, Ishikawa, & Silverstein, 1977)

Patterns were first applied to software in the 1980s by Ward Cunningham, Kent Beck, and Jim Coplien. Later, the famous "Gang of Four" book, *Design Patterns: Elements of Reusable Object-Oriented Software* (Gamma, Helm, Johnson, & Vlissides, 1995), popularized the use of patterns.

5.5.3   WHY DO WE NEED PATTERNS?

Developing software is hard and developing reusable software is even harder. Designs should be specific to the current problem, but general enough to address future problems and requirements. Experienced designers know not to solve every problem from first principles, but to reuse solutions encountered previously. They find recurring patterns and then use them as a basis for new designs. This is simply an embodiment of the Principle of Generality.

5.5.4 WHAT ARE THE BENEFITS OF PATTERNS?

First, design patterns help in finding appropriate objects, in determining object granularity, and in designing in anticipation of change. At the design level, design patterns enable large-scale reuse of software architectures by capturing expert knowledge and making this expertise more widely available. Finally, patterns help improve developer vocabulary (I am always amazed to listen to a group of software engineers with substantial pattern knowledge discuss designs in what seems like a different language). While the terminology might be unfamiliar to those not knowing the particular pattern language, the efficiency of information exchange is very high. Some people also contend that learning patterns can help in learning object-oriented technology.

5.5.5 WHAT DO PATTERNS LOOK LIKE?

In general, a pattern has four essential elements:

* A name
* A problem description
* A solution to the problem
* The consequences of the solution.

The name is simply a convenient handle for the pattern itself. Some of the names can be rather humorous (such as "flyweight" and "singleton"), but are intended to evoke an image to help remind the user of the intent of the pattern.

The problem part of the pattern template states when to apply the pattern; that is, it explains the problem and its context. The problem statement may describe specific design problems such as how to represent algorithms as objects. The problem statement may also describe class structures that are symptomatic of an inflexible design and possibly include conditions that must be met before it makes sense to apply the pattern.

The solution describes the elements that make up the design, although it does not describe a particular concrete design or implementation. Rather, the solution provides how a general arrangement of objects and classes solves the problem.

Finally, consequences show the results and trade-offs of applying pattern. It might include the impact of the pattern on space and time, language and implementation issues, and flexibility, extensibility, and portability. The consequences are critical for evaluating alternatives.

5.5.6 WHAT ARE THE "GRASP" PATTERNS?

The GRASP (general principles in assigning responsibilities) patterns are a fairly high-level set of patterns for design set forth by Craig Larman (2004). The GRASP patterns are:

- Creator
- Controller
- Expert
- Low coupling
- High cohesion
- Polymorphism
- Pure fabrication
- Indirected
- Protected variations.

Let's look at each of these patterns briefly, giving only name, problem, and solution in an abbreviated pattern template. The consequence of using each of these, generally, is a vast improvement in the OOD.

Name: Creator
Problem: Who should be responsible for creating a new instance of some class?
Solution: Assign Class B the responsibility to create an instance of class A if one or more of the following is true:
- B aggregates A objects
- B contains A objects
- B records instances of A objects
- B closely uses A objects.

B has the initializing data that will be passed to A when it is created.
   For example, in the baggage inspection system, the class "`camera`" would create objects from the class "`baggage_image`."

Name: Controller
Problem: Who should be responsible for handling an input system event?
Solution: Assign the responsibility for receiving or handling a system event message to a class that represents the overall system or a single use case scenario for that system.

In the baggage inspection system, the "`baggage_inspection_system`" class would be responsible for "`new_baggage`" events.

Name: Expert
Problem: What is a general principle of assigning responsibilities to objects?
Solution: Assign a responsibility to the information expert—the class that has the information necessary to fulfill the responsibility.

For example, in the baggage inspection system the class "`threat_detector`" would be responsible for identifying objects of the class baggage as being a possible threat.

Name: Low Coupling
Problem: How do we support low dependency, low change impact, and increased reuse?
Solution: Assign the responsibility so the coupling remains low.

Applying separation of concerns or Parnas partitioning principles would be helpful here.

Name: Polymorphism
Problem: How can we design for varying, similar cases?
Solution: Assign a polymorphic operation to the family of classes for which the cases vary.

For example, in the baggage inspection system we probably have different baggage object types (e.g., suitcase, golf club case, baby seat, etc.) and the algorithm for inspecting the images derived from each of these should be different. But the method that scans each image should be determined at run time depending upon the object, not through the use of a case statement.

Name: Pure fabrication
Problem: Where can we assign responsibility when the usual options based on expert lead to high coupling and low cohesion?
Solution: Create a "`behavior`" class whose name is not necessarily inspired by the domain vocabulary in order to reduce coupling and increase cohesion.

For example, in the baggage inspection system we might contrive a class "`unconventiaonl  luggage`" to describe any type of luggage that is nonconventional (e.g., a cardboard box, a laundry bag, or other unusual container).

Name: Indirection
Problem: How do we reduce coupling?
Solution: Assign a responsibility to an intermediate object to decouple two components.

Here we might create a "`second_look`" class to deal with baggage that might need re-imaging.

Name: Protected Variations
Problem: How can we design components so that the variability in these elements does not have an undesirable impact on other elements?
Solution: Identify points of likely variation or instability; assign responsibilities to create a stable interface around them.

This principle is essentially the same as information hiding and the open/closed principle.

TABLE 5.1
The Set of Design Patterns Popularized by the "Gang of Four"

Creational	Behavioral	Structural
Abstract factory	Chain of responsibility	Adapter
Builder	Command	Bridge
Factory method	Interpreter	Composite
Prototype	Iterator	Decorator
Singleton	Mediator	façade
	Memento	Flyweight
	Observer	Proxy
	State	
	Strategy	
	Template method	
	Visitor	

Source: Gamma et al., 1995.

5.5.7 WHAT ARE THE GANG OF FOUR PATTERNS?

This set of design patterns was first introduced by Gamma, Helm, Johnson, and Vlissides (the "Gang of Four" or "GoF") and popularized in a well-known text (Gamma et al., 1995). They describe 23 patterns organized by creational, behavioral, or structural intentions (Table 5.1).

Many of the GoF patterns perfectly illustrate the convenience of the name in suggesting the application for the pattern. For example, the flyweight pattern provides a design strategy when a large number of small-grained objects will be needed, such as baggage objects in the baggage inspection system. The singleton pattern is used when there will be a single instance of an object, such as the single instance of a baggage inspection system object. In fact, the singleton pattern can be used as the base class for any system object.

Explication of each pattern is beyond the scope of this text. Table 5.1 is provided for illustration only. Interested readers are encouraged to consult Gamma et al (1995). However, the three Gang of Four pattern types will now be discussed in general terms.

5.5.8 WHAT ARE CREATIONAL PATTERNS?

Creational patterns are connected with object creation, and they allow for the creation of objects without actually knowing what you are creating beyond the interfaces. Programming in terms of interfaces embodies information hiding. Therefore, we try to write as much as possible in terms of interfaces.

5.5.9 WHAT ARE STRUCTURAL PATTERNS?

Structural patterns are concerned with organization classes so that they can be changed without changing the other code. Structural patterns are "static model" patterns; that is, they represent structure that should not change.

5.5.10   WHAT ARE BEHAVIORAL PATTERNS?

Behavioral patterns are concerned with runtime (dynamic) behavior of the program. They help define the roles of objects and their interactions, but being dynamic they do not contain much, if any, structure

5.5.11   ARE THERE ANY OTHER PATTERN SETS?

There are many other pattern sets. For example, there is a well-known set of architecture and design patterns (Buschmann, Meunier, Rohnert, Sommerland, & Stal, 1996), analysis patterns (Fowler, 1996), and literally dozens of others.

5.5.12   WHAT ARE THE DRAWBACKS OF PATTERNS?

Patterns do not lead to direct code reuse. Direct code reuse is the subject of software libraries. Rather, patterns lead to reusable design and architectures, which can be converted to code.

Patterns are deceptively simple. While it might be easy enough to master the names of the patterns and memorize their structure visually, it is not so easy to see how they can lead to design solutions. This benefit takes a great deal of education and experience.

Teams may suffer from pattern overload, meaning that the quest to use pattern-based techniques can be an obsession and an end itself rather than the means to an end. Patterns are not a silver bullet. Rather, they provide another approach to solving design problems.

Finally, integrating patterns into a software development process is a labor-intensive activity; therefore, immediate benefits from a patterns program may not be realized.

5.6   DESIGN DOCUMENTATION

5.6.1   IS THERE A STANDARD FORMAT FOR SDS?

No. There are many different variations on the SDS (also called SDD). IEEE Standard 1016-2009, "IEEE Standard for Information Technology—Systems Design—Software Design Descriptions", is one possible resource that you can consult. But, every company uses a different template for this documentation.

5.6.2   HOW DO I ACHIEVE TRACEABILITY FROM REQUIREMENTS THROUGH DESIGN AND TESTING?

One way to achieve these links is through the use of an appropriate numbering system throughout the documentation. For example, a requirement numbered 3.2.2.1 would be linked to a design element with a similar number (the numbers don't have to be the same so long as the annotation in the document provides traceability). These linkages are depicted in Figure 5.8. Although the documents shown in the figure have not been

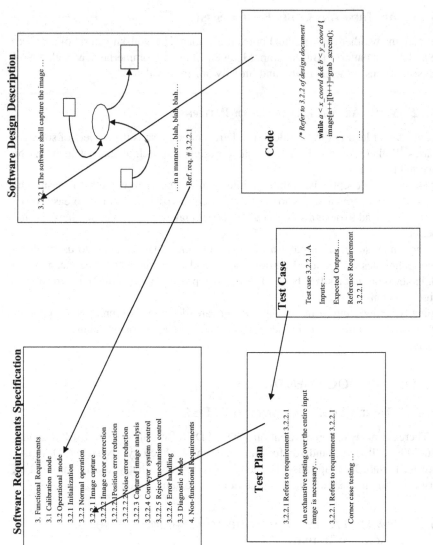

FIGURE 5.8 Linkages between software documentation and code. In this case, the links are achieved through similarities in numbering and specific reference to the related item in the appropriate document

TABLE 5.2
A Traceability Matrix Corresponding to Figure 5.8 Sorted by
Requirement Number

Requirement Number	SDD Reference Number	Test Plan Reference Number	Code Unit Name	Test Case Number
3.1.1.1	3.1.1.1	3.1.1.1	Simple_fun	3.1.1.A
	3.2.4	3.2.4.1		3.1.1.B
		3.2.4.3		
3.1.1.2	3.1.1.2	3.1.1.2	Kalman_filter	3.1.1.A
				3.1.1.B
3.1.1.3	3.1.1.3	3.1.1.3	Under_bar	3.1.1.A
				3.1.1.B
				3.1.1.C

introduced yet, the point is that the documents are all connected through appropriate referencing and notation.

Figure 5.8 is simply a graphical representation of the traceable links. In practice, a traceability matrix is constructed to help cross reference documentation and code elements (Table 5.2). The matrix is constructed by listing the relevant software documents and the code unit as columns, and then each software requirement as rows.

Constructing the matrix in a spreadsheet software package allows for providing multiple matrices sorted and cross-referenced by each column as needed. For example, a traceability matrix sorted by test case number would be an appropriate appendix to the text plan.

The traceability matrices are updated at each step in the software life cycle. For example, the column for the code unit names (e.g., procedure names, object class) would not be added until after the code is developed.

Finally, a way to foster traceability between code units is through the use of data dictionaries.

5.6.3   Can You Give an Example of a Design Document?

Appendix B contains an SDS for the wet well control system. Appendix C contains the object models for the wet well control system. The latter document is sometimes used as a supplement to an SDS.

FURTHER READING

Alexander, C., Ishikawa, S., & Silverstein, M. (1977). *A Pattern Language.* Oxford University Press, New York.

Beck, K., & Andres, C. (2004). *Extreme Programming Explained: Embrace Change.* 2nd edition. Addison-Wesley Professional.

Buschmann, F., Meunier, R., Rohnert, H., Sommerland, P., & Stal, M. (1996). *Pattern-Oriented Software Architecture, Volume 1: A System of Patterns.* John Wiley & Sons, New York.

Fowler, M. (1996). *Analysis Patterns: Reusable Object Models*. Addison-Wesley, Boston, MA.

France, R.B., Ghosh, S., Dinh-Trong, T., & Solberg, A. (2006). Model-driven development using UML 2.0: promises and pitfalls. *Computer, 39*(2), 59–66.

Gamma, E., Helm, R., Johnson, R., & Vlissides, J. (1995). *Design Patterns: Elements of Reusable Object-Oriented Software*. Addison-Wesley, Boston, MA.

Garlan, D. and Shaw, M. (January 1994). An Introduction to Software Architecture, Technical Report CMU/SEI-94-TR-21. Accessed from https://userweb.cs.txstate.edu/~rp31/papers/intro_softarch.pdf.

Kruchten, P., Obbink, H., & Stafford, J. (2006). The past, present, and future of software architecture. *IEEE Software, 23*(2), 22–30.

Laplante, P.A. (2004). *Software Engineering for Image Processing Systems*. CRC Press, Boca Raton, FL.

Larman, C. (2004). *Applying UML and Patterns*. Third Edition, Prentice Hall, Englewood Cliffs, NJ.

Levine, D. & Schmidt, D. (2000). *Introduction to Patterns and Frameworks*. Washington University, St Louis.

Liskov, B. & Wing, J. (1994). A behavioral notion of subtyping. *ACM Trans. Program. Lang. Syst., 16*(6), 1811–1841.

Martin, R.C. (2002). *Agile Software Development: Principles, Patterns, and Practices*. Prentice Hall, Englewood Cliffs, NJ.

Meyer, B. (2000). *Object-Oriented Software Construction*. 2nd ed., Prentice Hall, Englewood Cliffs, NJ.

Parnas, D.L. (1972). On the criteria to be used in decomposing systems into modules. *Commun. ACM, 15*(12), 1053–1058.

Pressman, R.S. & Maxim, B. (2019). *Software Engineering: A Practitioner's Approach*. 9th ed., McGraw-Hill, New York.

Sommerville, I. (2015). *Software Engineering*. 10th ed., Addison-Wesley, Boston, MA.

6 Software Construction

OUTLINE

- Programming languages
- Web-oriented programming
- Software construction tools
- Open-source software
- Becoming a better code developer

6.1 INTRODUCTION

When we talk of "constructing or building a software," we mean using tools to translate the designed model of the software system into runnable behavior using a programming language. The tools involved include the means for writing the code (a text editor), the means for compiling that code (the programming language compiler or interpreter), debuggers, and tools for managing the build of software on local machines and between cooperating computers across networks, including the Web.

Other tools for building software include version control and configuration management software and the integrated environments that put them together with the language compiler to provide for seamless code production. With these tools, the build process can be simple, especially when the specification, design, testing, and integration of the software are well-planned and executed.

In this chapter, we will discuss the software engineering aspects of programming languages and the software tools and practices used with them.

6.2 PROGRAMMING LANGUAGES

6.2.1 Is the Study of Programming Languages Really of Interest to the Software Engineer?

Yes. A programming language represents the nexus of design and structure. But a programming language is really just a tool and the best software developers are known for the quality of their tools and their skill with them. This skill can only be obtained through a deep understanding of the language itself and the peculiarities of a particular implementation of a language as seen in the compiler.

DOI: 10.1201/9781003218647-7

6.2.2 What Happens When Software Behaves Correctly but Is Poorly Written?

Misuse of the programming language is very often the cause of the reduction of the desirable properties of the software (such as maintainability and readability) and the increase in undesirable properties (such as fragility and viscosity).

6.2.3 But I Have Been Writing Code since My First Programming Course in College. Surely I Don't Need Any Lessons in Programming Languages, Do I?

Consider this rationale by Bruce Tate, from his book "Seven Languages in Seven Weeks: a pragmatic guide to learning programming languages.":

> As requirements for computer programs get more complex, languages, too, must evolve. Every twenty years or so, the old paradigms become inadequate to handle the new demands for organizing and expressing ideas. New paradigms must emerge, but the process is not a simple one.
>
> Each new programming paradigm ushers in a wave of programming languages, not just one. The initial language is often strikingly productive and wildly impractical. Think Smalltalk for objects or Lisp for functional languages. Then, languages from other paradigms build in features that allow people to absorb the new concepts while users can live safely within the old paradigm.
>
> (Tate, 2010)

6.2.4 What about Working with Legacy Code?

A great deal of legacy code is written in Fortran, COBOL, assembly language, Ada, Jovial, C, or any number of other exotic and antiquated languages. It is always hard to perform maintenance work on an old system implemented in one of these languages.

6.2.5 So How Many Programming Languages Are There?

There are literally hundreds of them, many of which are arcane or so highly specialized that there is no real benefit to discussing them. In the following, we discuss a few of the more frequently encountered ones.

6.2.6 Programming Language Landscape

6.2.6.1 What Does the Programming Landscape Look Like?

By focusing on general purpose, high-level programming languages. These languages can be analyzed as to their:

- Programming paradigm
- Their type system
- Whether they are compiled or interpreted.

6.2.6.2   What Are the Programming Paradigms?

A programming paradigm describes the way the language enables a programmer to achieve their goals. There are three programming paradigms:

- Procedural (e.g., COBOL, FORTRAN)
- Functional (e.g., Lisp, F#, Haskell)
- Object-oriented (e.g., Java, C++, C#).

These are not rigid categories because many languages support multiple paradigms. What is important is the predominant paradigm supported by the language.

In the procedural paradigm, the design is implemented in step-by-step instructions with limited support for abstraction. Functions are used as a way to break large operations into small operations. Also, there is a lot of mutable (changeable) state within a procedural language making it hard to protect data that, by design, should not change.

In the functional paradigm, the function *is* the design. Functions are first-class objects since they can be the value of an expression, passed as arguments, and put into data structures. By design, functional languages have good support for the immutable state since functions tend to be "pure," as they bar the assignment statement and avoid side effects.

Object-oriented programming languages are usually characterized by data abstraction, inheritance, and polymorphism. In the object-oriented paradigm, the design is implemented by collaborating objects (which are abstractions). Each object belongs to some class (which are also abstractions). If the class is well designed, the object is encapsulated so that it can evolve over time. The language may be sufficiently rich so it can support both mutable and immutable state for class data. Functions may be allowed in more contexts within the language (e.g. passed as parameters).

6.2.6.3   What Is Data Abstraction?

Data abstraction is the reduction of a particular body of data to a simplified representation of the whole. Abstraction, in general, is the process of taking away or removing characteristics from something in order to reduce it to a set of essential characteristics.

6.2.6.4   What Is Inheritance?

Inheritance allows the software engineer to define new objects in terms of previously defined objects so that the new objects "inherit" properties. For example, an object of type "bank account" has certain properties—e.g., it can be opened, closed, and deposits and withdrawals can be made. A "savings account" is a type of bank account which inherits those properties.

6.2.6.5   What Is Polymorphism?

Function polymorphism allows the programmer to define operations that behave differently, depending on the type of object involved. For example, a filter operation would act differently depending on the type of image and filtering needed. How the filter operation is applied is implemented at run time.

6.2.6.6 What Are Programming Languages Type Systems?

A data type is the set of allowed values that a language construct (expression, variable, function, class) can obtain. The type system of a language tries to regulate the interactions between all of the language constructs so as to limit bugs within a program. For example, the type system should prevent a string from being processed as a number and vice versa. A programming language is either statically typed or dynamically typed.

6.2.6.7 What Is a Statically Typed Language?

A statically typed language requires the declaration of a variable's type and consistent adherence to that type. Some type of coercion is allowed, but the coercion is well-documented and predictable. In an object-oriented language, downcasting is allowed, but this is considered dangerous. Due to this danger, type checking is performed by the runtime system. This keeps the entire language "type-safe."

Java is a statically typed language, for example. In recent years there has been much criticism of Java's verbosity and the constant need to repeat a variable's type (there have been advancements in type inferencing in recent releases). The upside of any statically typed language is that an integrated development environment (IDE) knows the type of any construct and can assist the programmer with auto-completion. This helps to avoid mistakes and improves productivity.

6.2.6.8 What Is a Dynamically Typed Language?

In contrast with static typing, there is dynamic typing. We define dynamic typing as the ability to add methods and properties to a class at run time. A dynamically typed language is very good for creating frameworks and domain-specific languages. This ability is a productivity boon for the particular application area.

The downside of a dynamically typed language is that errors like misspellings cannot be detected at compile time. Such errors are caught at run time. JavaScript is such a language. If you have ever had to develop and debug a large JavaScript program, you know how frustrating the language can be.

6.2.6.9 What Is the Difference between Compiled and Interpreted Languages?

Another dimension along which languages can be compared is the compiled vs. interpreted. A compiled language is easy to recognize due to the distinct compilation phase that precedes the execution of the program. An interpreted language lacks this phase. An interpreter processes the source code, commonly called a script, immediately and thus runs the program.

A compiled language results in faster execution of the program because the compiler has handled at least half of the work—lexical analysis and syntax checking. Also, compiled programs can be processed by a number of optimizers to produce highly efficient code.

TABLE 6.1
Language Comparison

Language	Paradigm	Type System	Compiled vs. Interpreted
Java	OO w/ some Functional	Static	Compiled
JavaScript	Some OO	Dynamic	Interpreted
	Some Functional		
Groovy	OO w/ Some Functional	Dynamic	Compiled
Scala	OO and Functional	Static	Compiled

An interpreter must process a script on the fly. It must perform lexical analysis, syntax validation, and translation to some kind of internal representation. Since it has to do so much, it has little opportunity for optimizing the code. One advantage of an interpreted language, however, is that it can identify potential syntax errors as you are typing in the code, rather than during compile when many accumulated errors can appear.

6.2.6.10 How Do I Classify Programming Languages according to This Scheme (Programming Paradigm, Type System, and Whether They Are Compiled or Interpreted)?

To make the discussion more concrete, we will compare and contrast some purposely chosen languages: Java, JavaScript, Groovy, and Scala. We assume that the reader is somewhat familiar with Java and JavaScript. Groovy is a dynamic version of Java and is directly comparable to Ruby. Scala is a hybrid functional and object-oriented language. It is a type-safe language with appealing concurrency features. Interestingly, each of these languages runs on the Java Virtual Machine. Table 6.1 indicates the primary features of these languages.

6.2.7 PROGRAMMING FEATURES AND EVALUATION

6.2.7.1 Which Programming Language Is Better?

Unfortunately, the answer is, "it depends." It is like the question "Which is better, Italian food or Chinese food?" You can't make objective comparisons because it depends upon whom you ask. So, it is with programming languages.

Now, some programming languages are better for certain applications or situations; for example, C is well suited for certain embedded systems, scripting languages are better suited for rapid prototyping, and so forth. Indeed, each programming language offers its own strengths and weaknesses with respect to specific applications domains.

6.2.7.2 So, What Is the Best Way to Evaluate a Programming Language?

We like Cardelli's (1996) evaluation criteria. He classifies languages along the following dimensions.

- The economy of execution: How fast does the program run?
- The economy of compilation: How long does it take to go from sources to executables?
- The economy of small-scale development: How hard must an individual programmer work?
- The economy of large-scale development: How hard must a team of programmers work?
- The economy of language features: How hard is it to learn or use a programming language?

But, as discussions around these dimensions can get somewhat theoretical, we will use more tangible "visible" features to discuss the more commonly encountered programming languages.

6.2.7.3   What Do You Mean by Visible Features of Programming Languages?

There are several programming language features that stand out that help promote the desirable properties of software and best engineering practices. These are as follows:

- Versatile parameter passing mechanisms
- Dynamic memory allocation facilities
- Strong typing
- Abstract data typing
- Exception handling
- Modularity

6.2.7.4   Why Should I Care about Parameter-Passing Techniques?

The use of parameter lists is likely to promote modular design because the interfaces between the modules are clearly defined. There are several methods of parameter passing, but the three most commonly encountered are call-by-value, call-by-reference, and global variables.

Clearly defined interfaces can reduce the potential of untraceable corruption of data by procedures using global access. However, both call-by-value and call-by-reference parameter-passing techniques can impact performance when the lists are long because interrupts are frequently disabled during parameter passing to preserve the time correlation of the data passed. Moreover, call-by-reference can introduce subtle function side effects that depend on the compiler.

6.2.7.5   What Is Call-by-Reference?

In call-by-reference or call-by-address, the address of the parameter is passed by the calling routine to the called procedure so that it can be altered there. Execution of a procedure using call-by-reference can take longer than one using call-by-value because indirect mode instructions are needed for any calculations involving the variables passed in call-by-reference. However, in the case of passing large data structures between procedures, it is more desirable to use call-by-reference because passing a pointer to a large data structure is more efficient than passing the structure field by field.

6.2.7.6 What Is Call-by-Value?

In call-by-value parameter passing, the value of the actual parameter in the subroutine or function call is copied into the formal parameter of the procedure. Because the procedure manipulates the formal parameter, the actual parameter is not altered. This technique is useful when either a test is being performed or the output is a function of the input parameters. For example, in an edge detection algorithm, an image is passed to the procedure and some description of the location of the edges is returned, but the image itself need not be changed.

When parameters are passed using call-by-value, they are copied onto a runtime stack at a considerable execution time cost. For example, large data structures must be passed field by field.

6.2.7.7 What about Global Variables?

Global variables are variables that are within the scope of all modules of the software system. This usually means that references to these variables can be made in direct mode and thus are faster than references to variables passed via parameter lists. For example, in many image processing applications, global arrays are defined to represent images. Hence, costly parameter passing can be avoided.

Global variables can be dangerous because the reference to them may be made by unauthorized code, thus introducing subtle bugs. For this and other reasons, the unwarranted use of global variables is to be avoided. Global parameter passing is recommended only when timing warrants and its use must be clearly documented.

6.2.7.8 How Do I Choose Which Parameter-Passing Technique to Use?

The decision to use one method of parameter passing or the other represents a trade-off between good software engineering practice and performance needs. For example, often timing constraints force the use of global parameter passing in instances when parameter lists would have been preferred for clarity and maintainability.

6.2.7.9 What Is Recursion?

Most programming languages provide a recursion feature which is when a procedure can call itself. For example, in pseudo-code:

```
foobar(int x, int y)
{
  if
  (x < y) foobar(y-x, x); else
    return (x);
}
```

Here, foobar is a recursive procedure involving two integers. Invoking foobar(1,2) will return the value 1.

Recursion is widely used in many mathematical algorithms that underlie engineering applications. Recursion can simplify the programming of non-engineering applications as well.

6.2.7.10   Are There Any Drawbacks to Recursive Algorithm Formulations?

Yes. While recursion is elegant and often necessary, its adverse impact on performance must be considered. Procedure calls require the allocation of storage on one or more stacks for the passing of parameters and for storage of local variables.

The execution time needed for the allocation and deallocation, and for the storage of those parameters and local variables can be costly. In addition, recursion necessitates the use of a large number of memory access events which slow processing.

Finally, the use of recursion often makes it impossible to determine the size of runtime memory requirements. Thus, iterative techniques such as **while** and **for** loops must be used if performance prediction is crucial or in those languages that do not support recursion.

6.2.7.11   What Does "Dynamic Memory Allocation" Mean?

This means that memory storage requirements need to be known before the program is written so that memory can be allocated and deallocated as needed. The capability to dynamically allocate memory is important in the construction and maintenance of many data structures needed in a complex engineering system. While dynamic allocation can be time-consuming, it is usually necessary, especially when creating intermediate data structures needed in many engineering algorithms.

6.2.7.12   What Are Some Examples of Dynamic Allocation Use?

Linked lists, trees, heaps, and other dynamic data structures can benefit from the clarity and economy introduced by dynamic allocation. Furthermore, in cases where just a pointer is used to pass a data structure, then the overhead for dynamic allocation can be quite reasonable. When writing such code, however, care should be taken to ensure that the compiler will pass pointers to large data structures and not the data structure itself.

6.2.7.13   What Is Meant by "Strong Typing" in a Programming Language?

Strongly typed languages require that each variable and constant be of a specific type (e.g., float, Boolean, or integer) and that each be declared as such before use. Generally, high-level languages provide integer and floating-point types, along with Boolean, character, and string types. In some cases, abstract data types are supported. These allow programmers to define their own type along with the associated operations.

Strongly typed languages prohibit the mixing of different variable types in operations and assignments, and thus force the programmer to be precise about the way data are to be handled. Precise typing can prevent corruption of data through unwanted or unnecessary type conversion.

Weakly typed languages either do not require explicit variable type declaration before use (those of you familiar with old versions of Fortran or BASIC will recognize this concept) or do not prohibit the mixing of types in arithmetic operations. Because these languages generally perform mixed calculations using the type that has the highest storage complexity, they must promote all variables to that type. For example, in C, the following code fragment illustrates the automatic promotion and demotion of variable types:

```
int x,y; float k,l,m;
 .
 .
 .
j = x*k+m;
```

Here the variable x will be promoted to a float (real) type and then multiplication and addition will take place in the floating-point. Afterward, the result will be truncated and stored in j. The performance impact is that hidden promotion and more time-consuming arithmetic instructions can be generated with no additional accuracy. In addition, accuracy can be lost due to the truncation, or, worse, an integer overflow can occur if the floating-point value is larger than the allowable integer value.

Programs written in languages that are weakly typed need to be scrutinized for such effects. Some C compilers will catch type mismatches in function parameters. This can prevent unwanted type conversions.

6.2.7.14 What Is Exception Handling?

Certain languages provide facilities for dealing with errors or other anomalous conditions that arise during program execution. These conditions include the obvious, such as floating-point overflow, square root of a negative, divide-by-zero, and image-related conditions such as boundary violation, wraparound, and pixel overflow.

The capability to define and handle exceptional conditions in high-level languages aids in the construction of interrupt handlers and other code used for real-time event processing. Moreover, poor handling of exceptions can degrade performance. For example, floating-point overflow errors can propagate bad data through an algorithm and instigate time-consuming error recovery routines.

6.2.7.15 Which Languages Have the Best Exception Handling Facilities?

Java has excellent exception handling through its "try, catch, finally" approach, which is used in many mainstream object-oriented languages. In Java, the structure looks something like this:

```
try
{
    // do something
}
catch (Exception1)
{
    // what to do if something goes wrong
}
catch (Exception2)
{
```

```
       // what to do if something else goes wrong
    }
    ...
    finally
    {
       // what to do if all else fails
    }
```

In this way, foreseeable "risky" computation can be handled appropriately, rather than relying on the operating system to use its standard error handling, which may be inadequate.

Ada95, C++, and C# also have excellent exception handling capabilities. ANSI C provides some exception handling capability through the use of signals.

6.2.7.16   What Is Meant by Modularity?

Procedural languages that are amenable to the principle of information hiding and separation of concerns tend to make it easy to construct subprograms. While C and Fortran both have mechanisms for this (procedures and subroutines), other languages such as Ada95 (which can be considered either procedural or object-oriented) tend to foster more modular design because of the requirement to have clearly defined input and outputs in the module parameter lists.

In Ada, the notion of a package exquisitely embodies the concept of Parnas information hiding. The Ada package consists of a specification and declarations that include its public or visible interface and its invisible or private parts. In addition, the package body, which has further externally invisible components, contains the working code of the package. Packages are separately compilable entities, which further enhances their application as black boxes.

In Fortran, there is the notion of a subroutine and separate compilation of source files. These language features can be used to achieve modularity and design abstract data types.

The C language also provides for separately compiled modules and other features that promote a rigorous top-down design approach, which should lead to a good modular design.

While modular software is desirable, there is a price to pay in the overhead associated with procedure calls and parameter passing. This adverse effect should be considered when sizing modules.

6.2.7.17   Do Object-Oriented Languages Support a Form of Modularity?

Object-oriented languages provide a natural environment for information hiding. For example, in image processing systems, it might be useful to define a class of type pixel, with attributes describing its position, color, and brightness; and operations that can be applied to a pixel such as add, activate, deactivate, and so on. It might also be desirable to define objects of type image as a collection of pixels with other attributes of width, height, and so on.

In some cases, expression of system functionality is easier to do in an object-oriented manner.

6.2.7.18 What Is the Benefit of Object Orientation from a Programming Perspective?

Object-oriented techniques can increase programmer efficiency, reliability, and the potential for reuse. More can be said on this subject, but the reader is referred to the references at the end of the chapter (Gamma et al., 1995; Meyer, 2000).

6.2.8 BRIEF SURVEY OF LANGUAGES

6.2.8.1 Can You Apply the Micro Properties Just Discussed to Some of the More Commonly Used Programming Languages?

For purposes of illustrating the aforementioned language properties, let us review some of the more widely used languages in engineering systems.

- C
- C++
- C#
- Fortran
- Java
- Rust
- Java Script
- Python
- Perl.

Functional languages, such as Lisp and ML, have been omitted from the discussions. This is simply because their use in this setting is rare.

6.2.8.2 When Can C Be Used?

The C programming language, invented around 1971, is a good language for "low-level" programming. This is because C is descended from the language BCPL (whose successor, C's parent, was "B"), which supported only one type—machine word. Consequently, C supported machine-related objects like characters, bytes, bits, and addresses, which could be handled directly in a high-level language. These entities can be manipulated to control interrupt controllers, CPU registers, and other hardware needed in embedded systems. C is sometimes used as a high-level cross-platform assembly language. The C language is good for embedded programming because it provides structure and flexibility without complex language restrictions.

6.2.8.3 What Other Useful Features Does C Have?

C provides special variable types such as register, volatile, static, and constant, which allow for control of code generation at the high-order language level. For example, declaring a variable as a register type indicates that it will be used frequently. This encourages the compiler to place such a declared variable in a register, which often results in smaller and faster programs. C supports call-by-value only, but call-by-reference can be implemented by passing the pointer to anything as a value.

Variables declared as type volatile are not optimized by the compiler. This is useful in handling memory-mapped Input/Output (I/O) and other instances where the code should not be optimized.

6.2.8.4   How Does C Handle Exceptions?

The C language provides for exception handling through the use of signals, and two other mechanisms, setjmp and longjmp, are provided to allow a procedure to return quickly from a deep level of nesting. This is a useful feature in procedures requiring an abort. The setjmp procedures call, which is really a macro (but often implemented as a function), saves environment information that can be used by a subsequent longjmp library function call. The longjmp call restores the program to the state at the time of the last setjmp call. Procedure process is called to perform some processing and error checking. If an error is detected, a longjmp is performed, which changes the flow of execution directly to the first statement after the setjmp.

6.2.8.5   What Is the Relationship between C and C++?

C++ is a hybrid object-oriented programming language that was originally implemented as a macro-extension of C. Today, C++ stands by itself as a separately compiled language, although strictly speaking, C++ compilers should accept standard C code.

C++ exhibits all three characteristics of an object-oriented language. It promotes better software engineering practice through encapsulation and better abstraction mechanisms, such as inheritance, composition, and polymorphism, than does C.

6.2.8.6   When Should C++ Be Used?

Significantly, more embedded systems are being constructed in C++ and many practitioners are asking this question. My answer to them is always "it depends." Choosing C in lieu of C++ in embedded applications is, roughly speaking, a tradeoff between a "lean and mean" C program that will be faster and easier to predict but harder to maintain and a C++ program that will be slower and unpredictable but potentially easier to maintain.

C++ still allows for low-level control (and not falling back to C features). For example, it can use in-line methods rather than a runtime call. This kind of implementation is not completely abstract, nor completely low-level, but is acceptable in embedded environments.

6.2.8.7   What Is the Danger in Converting My C Code to C++?

There is some tendency to take existing C code and "objectify" it by wrapping the procedural code into objects with little regard for the best practices of object orientation. This kind of approach should be avoided because it has the potential to incorporate all of the disadvantages of C++ and none of the benefits.

6.2.8.8   Can You Tell Me about C#?

C# (pronounced C Sharp) is an object-oriented programming language released in 2002 by Microsoft and stands today as a much-loved improvement on the C++ coding language. It is mainly being used for game development, desktop/web/mobile apps, and virtual reality applications. Some notable features of C# are:

- Unlike C++, C# does not support multiple inheritances, although a class can implement any number of "interfaces" (fully abstract classes). This was a design decision by the language's lead architect to avoid complications and simplify architectural requirements throughout Common Language Infrastructure.
- Exception handling provides a structured and extensible approach to error detection and recovery.
- Garbage collection automatically reclaims memory occupied by unreachable unused objects.
- Lambda expressions support functional programming techniques.
- Language-Integrated Query (LINQ) syntax creates a common pattern for working with data from any source.
- Language support for asynchronous operations provides a syntax for building distributed systems.
- C# supports strongly, implicitly typed variable declarations with the keyword `var`, and implicitly typed arrays with the keyword `new[]` followed by a collection initializer. Nullable types guard against variables that don't refer to allocated objects.

6.2.8.9   Can You Tell Me about Fortran?

Fortran is the oldest high-order language (developed circa 1955) but its modern variants are still in regular use today, particularly in certain scientific and engineering applications. Because in its earlier versions it lacked recursion and dynamic allocation facilities, more complex systems written in this language typically included a large portion of assembly language code to handle interrupts and scheduling. Communication with external devices was by memory-mapped I/O, Direct Memory Access (DMA), and I/O instructions. Later versions of the language included features such as reentrant code, but even today a complex Fortran control system may require some assembly language code to accompany it.

To its detriment, Fortran is weakly typed, but because of the subroutine construct and the `if-then-else` construct, it can be used to design highly structured code. Fortran has no built-in exception handling or abstract data types.

Today, Fortran is still used to write some engineering applications because of its excellent handling of mathematical processing and because "old-time" engineers learned this language first. There is even a "new" language called F, which is a derivative of Fortran, less some esoteric and dangerous features. Many legacy systems, particularly in engineering applications, can still be found to have been written in "plain old" Fortran.

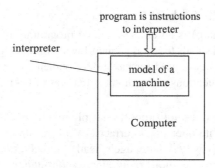

FIGURE 6.1   The Java interpreter as a model of a virtual machine

6.2.8.10   Can You Tell Me about Java?

Java is an interpreted language; that is, the code compiles into machine-independent code which runs in a managed execution environment. This environment is a virtual machine (Figure 6.1), which executes "object" code instructions as a series of program directives.

The advantage of this arrangement is that the Java code can run on any device that implements the virtual machine. This "write once, run anywhere" philosophy has important applications in embedded and portable computing as well as in web-based computing.

6.2.8.11   If Java Is Interpreted, Does That Mean It Cannot Be Compiled?

There are native-code Java compilers that allow Java to run directly "on the bare metal;" that is, the compilers convert Java directly to assembly code or object code.

6.2.8.12   What Are Some of the Main Features of Java?

Java is an object-oriented language that looks very much like C++. Like C, Java supports call-by-value, but the value is the reference to an object, which is in essence call-by-reference for all objects. Primitives are passed-by-value.

6.2.8.13   Are There Any Well-Known Problems with Using Java?

One of the best-known problems with Java involves its garbage collection utility. Garbage is a memory that has been allocated but is unusable because of the loss of a pointer to it; for example, through the destruction of an object. The allocated memory must be reclaimed through garbage collection.

Garbage collection algorithms generally have unpredictable performance (although average performance may be known). The loss of determinism results from the unknown amount of garbage, the tagging time of the non-deterministic data structures, and the fact that many incremental garbage collectors require that every memory allocation or deallocation from the heap be willing to service a page-fault trap handler.

6.2.8.14   What Are the Differences between Java and C++?

The following are some features of Java that are different from C++ and that are of interest in engineering applications:

* There are no global functions or constants — everything belongs to a class.
* Arrays and strings have built-in bounds checking.
* All values are initialized and use special defaults if none are given.
* All classes ultimately inherit from the object class.
* Java does not support the goto statement. However, it supports labeled breaks, which can be used in the same way.
* Java does not support automatic type conversions (except where guaranteed safe).
* Types are all references to objects, except the primitive types (e.g., integer, floating-point, and Boolean).

6.2.8.15   What Are the Differences between Java and C#?

While Java runs on the Java Runtime Environment (JRE), C# is designed to be run on the Common Language Runtime. Java is a class-based object-oriented language whereas C# is a multi-paradigm language (object-oriented, functional, strong typing, component-oriented). Java does not support pointers while C# supports pointers only in an unsafe mode. Java doesn't support operator overloading whereas C# provides operator overloading for multiple operators.

6.2.8.16   Can You Tell Me about Rust Programming Language?

Rust first appeared in 2010 as a general-purpose programming language designed for performance and safety, especially safe concurrency. Similar to C#, Rust is a multi-paradigm language. Rust achieves memory safety without garbage collection. Rust has been called a systems programming language and, in addition to high-level features such as functional programming, it also offers mechanisms for low-level memory management.

6.2.8.17   What about Legacy Code Written in Arcane Languages Such as BASIC, COBOL, and Scheme?

Code in many arcane languages still abounds. It is hard to explain how to deal with these situations except on a case-by-case basis.

6.2.8.18   What about Visual Basic?

There are different versions of Visual Basic, the oldest being derived from and resembling in many ways the ancient BASIC language. The newer versions exhibit strong object-oriented features. A great deal of production code has been written in Visual Basic, but as this is a proprietary language, it is beyond the scope of this text. Of course, there are many good texts available in this language.

6.2.8.19   What about Scripting Languages Like JavaScript, Perl, and Python?

Object-oriented scripting languages, such as JavaScript, Perl, Python, and Ruby, have become quite popular. Most scripting languages make it easier to write programs because of weak typing and their interpreted nature. The interpreted code leads to a shorter build cycle (high economy of compilation), which, in turn, allows for rapid prototyping and exploratory changes, a feature that is useful in agile development methodologies.

6.3   WEB-ORIENTED PROGRAMMING

6.3.1   What Is Web-Oriented Programming?

Web programming refers to the writing, markup, and coding involved in Web development, which includes content, client and server scripting, and connection security. Web-oriented Programming (WOP) is:

- HTTP based
- Platform-independent
- Service-oriented toward program-to-program communication
- Not fully based on standards

So does the fact that a WOP application is platform-independent mean that the hardware and the programming language are not relevant? No—what is important is that the application provides a service to another program. The program may have a user interface (UI) or it may just provide a programming interface.

6.3.2   What Is the Role of the HTTP Protocol in the WOP?

The HTTP protocol is fundamental to WOP. The fundamentals of HTTP-based client/server applications are based on the browser/web server model where the application requests a resource (document) and the web server returns an HTML file to be interpreted by the browser.

The HTTP protocol has a number of request methods (GET, POST, PUT, DELETE) so that the user is not restricted to just "reading" what the server has to send. The HTTP protocol is mature in the sense that each of these verbs can have request parameters that further specify the type of response that the client can process.

Another characteristic of HTTP is that it is inherently stateless. The server can remain unaware of who the client is and what was returned to the client. This enables caching of results and caching enables scaling of the application.

The early days of the Internet were chaotic. Standards and protocols were both immature and generally not followed. Just as early browsers accepted bad HTML, web servers accepted "impure" GET and POST requests.

Web developers and systems integrators were looking for a better, standards-based approach to providing robust web services over the Internet. This leads to SOAP and its related technologies.

6.3.3 WHAT IS SOAP?

The Standard Object Access Protocol or SOAP is a messaging protocol specification for exchanging structured information in the implementation of web services in computer networks. In order to understand SOAP, you need to appreciate the problem it was designed to solve. The problem is automating an exchange of goods or information between a consumer and a provider, both of which are programs.

The SOAP architects designed the following components (Figure 6.2):

- Registry: known URI (Uniform Resource Identifier) where participating service providers register their services
- Provider: provider of a web service (a program)
- Consumer: consumer of a web service (a program).

Each component resides on a different node with different ownership. Communication between nodes uses the HTTP protocol. Figure 6.2 shows the message interaction between these components.

The labeled transitions are described as:

1. A service provider describes its service using WSDL (Web Services Description Language). This definition is published in a repository of services. The repository could use Universal Description, Discovery, and Integration (UDDI). Other forms of directories could also be used.
2. A service consumer issues one or more queries to the repository to locate a service and determine how to communicate with that service.
3. Part of the WSDL provided by the service provider is passed to the service consumer. This tells the service consumer the requests and responses of the service provider.

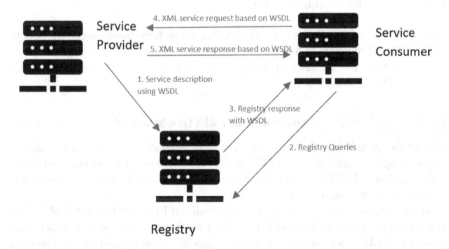

FIGURE 6.2   SOAP model for web services

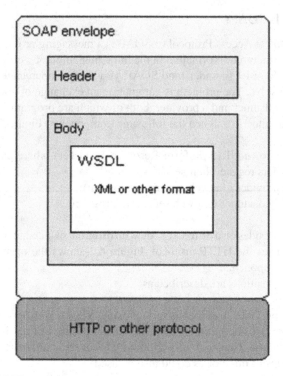

FIGURE 6.3   SOAP message

4. The service consumer uses the WSDL to send a request to the service provider.
5. The service provider provides the expected response to the service consumer.

SOAP provides the envelope for sending messages over the Internet. The SOAP envelope contains two parts (Figure 6.3):

1. An optional header provides information on authentication, encoding of data, or how a recipient of a SOAP message should process the message.
2. The body that contains the message. The messages can be defined using the WSDL specification.

6.3.4   Can You Say Something about RESTful Services?

Due to the complexity of SOAP and XML, an alternative to "standard" web services emerged based on pure HTTP verbs. This alternative has no standard and is simply called "RESTful services" or a "RESTful API". REST is an abbreviation of Representational State Transfer.

A RESTful service is exposed through a Uniform Resource Locator (URL). This is a logical name that separates the identity of the resource from what is accepted or returned. The URL acts as a handle for the resource, something that can be requested, deleted, or updated.

There are three key aspects of RESTful services. The first is that each of the resources on the server must have an identifier so that the request is unambiguous as to its reference. The second is that the client of such a service may specify what type of response it prefers by specifying an "Accept" header in the request. If the server supports that representation, it will return it. Finally, each request must contain enough state to answer the request. Together these aspects allow for statelessness on the server, and thus enable scaling the server and caching of specific requests.

6.4 SOFTWARE CONSTRUCTION TOOLS

6.4.1 WHAT IS THE VALUE PROPOSITION FOR USING SOFTWARE CONSTRUCTION TOOLS?

Management of the software development phase can be greatly improved with the usage of the proper tools such as version control tools that regulate access to the various components of the system from the software library, CASE tools assist with software construction, and supporting many other software engineering activities. In this section, we provide a high-level discussion of some of the software construction tools that you shall be familiar with.

6.4.2 WHAT IS A COMPILER?

A compiler translates a program from high-level source code language into relocatable object code or machine instructions or into an intermediate form such as an assembly language (a symbolic form of machine instructions). This process is broken into a series of phases, which will be described shortly.

After compilation, the relocatable code output from the compiler is made absolute, and a program called a linker or linking loader resolves all external references. Once the program has been processed by the linker, it is ready for execution. The overall process is illustrated in Figure 6.4.

This process is usually hidden by any integrated development environment that acts as a front end to the compiler.

6.4.3 CAN YOU DESCRIBE FURTHER THE COMPILATION PROCESS?

Consider the widely used Unix-based C compiler. The Unix C compiler **cc** provides a utility that controls the compilation and linking processes. This compiler is also found in many Linux implementations. In particular, in Unix System V version 3 (SVR3), the **cc** program provides for the following phases of compilation:

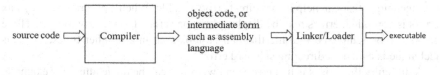

FIGURE 6.4 The compilation and linking process

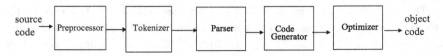

FIGURE 6.5   Phases of compilation provided by a typical compiler, such as C

TABLE 6.2
Phases of Compilation and Their Associated Program for a C Compiler

Phase	Program
Preprocessing	cpp
Compilation	ccom
Optimization	c2
Assembly	as
Linking and loading	ld

- Preprocessing
- Compilation
- Optimization
- Assembly
- Linking and loading.

These phases are illustrated in Figure 6.5 and summarized in Table 6.2. The preprocessing phase of the compiler, performed by the program **cpp**, takes care of things such as converting symbolic values into actual values and evaluating and expanding code macros. The compilation of the program, that is, the translation of the program from C language to assembly language, is performed by the program **ccom**.

Optimization of the code is performed by the program **c2**, and assembly or translation of the assembly language code into machine code is taken care of by **as**. Finally, the object modules are linked together, all external references (e.g., library routines) are resolved, and the program (or an image of it) is loaded into memory by the program **ld**. The executable program is now ready to run.

The fact that many of these phases can be bypassed or run alone is an important feature that can help in program debugging and optimization.

6.4.4   How Do I Deal with Compiler Warnings and Errors?

One technique that can help is to redirect errors to a file. When a program with syntax errors is compiled, errors may be displayed on the screen too quickly to read. These errors can be redirected to a file that can be looked at in a steadier timescale. This technique is called "redirecting standard error."

A difficulty that arises is that errors and warnings can be misleading; for example, an error reported in one place is indicative of a problem somewhere else. With experience, an error message can be associated with a particular problem, even when the error message appears far away from the error itself.

Another difficulty with errors and warnings is the cascade effect—the compiler finds one error and because no recovery is possible, many other errors follow. Again, experience with a particular compiler will enable the software engineer to more easily work through the cascade of errors and find the root cause.

It is beyond the scope of this text to discuss the many kinds of program warnings and errors that the user can encounter in the course of compiling and linking programs. A complete discussion of this issue is best left to the reference book of the language in question.

6.4.5   Do You Have Any Debugging Tips?

Programs can be affected by syntactic or logic errors. Syntactic or syntax errors arise from the failure to satisfy the rules of the language. A good compiler will always detect syntax errors, although the way that it reports the error can often be misleading.

For example, in a C program a missing curly brace (}) may not be detected until many lines after it should have appeared, but in deeply nested arrays of curly braces it may be hard to see where the missing brace belongs.

In logic errors, the code adheres to the rules of the language but the algorithm that is specified is somehow wrong. Logic errors are more difficult to diagnose because the compiler cannot detect them. A few basic rules may help you find and eliminate logic errors.

- Document the program carefully. Ideally, each nontrivial line of code should include a comment. In the course of commenting, this may detect or prevent logical errors.
- Where symbolic debugging is available, use steps, traces, break-points, skips, and so on, to isolate the logic error (discussed later).
- In the case of a command-line environment (such as Unix/Linux), output intermediate results at checkpoints in the code. This may help detect logic errors.
- In the case of an error, comment out portions of the code until the program compiles and runs. Add in the commented out code, one feature at a time, checking to see that the program still compiles and runs. When the program either does not compile or runs incorrectly, the last code you added is involved in the logic error.

Finding and eliminating logic errors is more art than science, and the software engineer develops these skills only with time and practice. In many cases, code audits or walkthroughs can be quite helpful in finding logic errors.

6.4.6   Is There Any Way to Automatically Debug Code?

It is impossible to provide automatic logic validation and the compiler can check only for syntactical correctness. Many programming environments provide tools that are helpful in eliminating logical errors. For example, two tools (**lint** and **cb**) are associated with Unix and Linux. As its name implies, **lint** is a nit-picker that does checking beyond that of an ordinary compiler. For example, C compilers are often not particular about certain inconsistencies such as parameter mismatches, declared variables that are not used, and type checking—**lint** is, however. Often, very difficult bugs can be prevented or diagnosed by using **lint**.

The C beautifier, or **cb**, is used to transform a sloppy-looking program into a readable one. The program code is not changed by **cb**. Instead, it just adds tabs, line feeds, and spaces where needed to make things look nice. This is very helpful in finding badly matched or missing curly braces, erroneous if-then-else and case statements, and incorrectly terminated functions. As with **lint**, **cb** is run by typing **cb** and a file name at the command prompt.

Many open-source integrated development environments (IDEs) have plug-ins that are the equivalent of or better than **lint** and **cb** and there are many other tools available, such as automatic refactoring engines, that can help improve code. Open-source software is discussed later in this chapter.

6.4.7   What Are Symbolic Debuggers and How Are They Used?

Symbolic, or source-level, debuggers are software programs that provide the ability to step through code at either a macro-assembly or high-order language level. They are extremely useful in module-level testing. They are less useful in system-level debugging because the real-time aspect of the system is necessarily disabled or affected.

Debuggers can be obtained as part of compiler support packages or in conjunction with sophisticated logic analyzers. For example, **sdb** is a generic name for a symbolic debugger associated with Unix and Linux. It is a debugger that allows the engineer to single-step through source code language and view the results of each step. GNU has the GNU debugger, **gdb**, which can be used with GNU C and other languages such as GNU Java and Fortran.

In order to use the symbolic debugger, the source code must be compiled with a particular option set. This has the effect of including special runtime code that interacts with the debugger. Once the code has been compiled for debugging, then it can be executed "normally."

6.4.8   Can You Give Me an Example of a Debugging Session?

In the Unix/Linux environment, the program can be started normally from the **sdb** debugger at any point by typing certain commands at the command prompt. However, it is more useful to single-step through the source code. Lines of code are displayed and executed one at a time by using the "s" (for step) command. If the statement is an output statement, it will output to the screen accordingly. If the statement is an input statement, it will await user input. All other statements execute normally. At any point in the single-stepping process, individual variables can be set or examined.

There are many other features of **sdb**, such as breakpoint setting. In more integrated development environments, a graphical user interface (GUI) is also provided, but these tools essentially provide the same functionality.

6.4.9 What Is a Version Control System?

This is the system responsible for managing changes to computer programs, documents, large websites, or other collections of information. Version control is a component of software configuration management.

6.4.10 What Are the Benefits of Using VCS?

VCS benefits software developers because it provides a:

1. Backup mechanism
2. Time machine (for retrieving past versions of the software)
3. Sharing mechanism.

A VCS is an enhanced backup system that allows you to go back in the history of a project. You can go back an hour, a day, or a year and know the exact state of the project at that time.

Not only can you go back in time, a VCS allows you to compare two previous versions of a file at different points in time. This is extremely useful if you are trying to track the point at which a defect was introduced into a project.

And, as a sharing mechanism, the VCS provides features that deter simultaneous access to the same file and also allows for tracking of who made what changes and when.

A VCS also guarantees that there is a copy of your intellectual property, i.e., your source code.

For these reasons alone, a VCS will benefit a solo developer who manages a code base. But a VCS shines as a team development tool because it allows multiple developers to make concurrent changes to the same set of files while maintaining the integrity of those files.

Version Control Systems date back to the implementations provided by system houses (IBM, Burroughs, Sperry, etc.). The first widely used system was SCCS (Source Code Control System) on early Unix systems. Today there are numerous commercial and open-source VCSs.

There are two types of VCS systems: centralized and distributed.

6.4.11 What Is a Centralized VCS?

A centralized VCS has a star topology with the shared repository at the center (Figure 6.6). The main feature of a centralized VCS is its simplicity. There is only one repository and that is where the "golden" copy of the source is be found. The usage commands are also simple: 1) commit my changes to the repository; 2) update my working copy with changes from other team members.

6.4.12 What Is a Distributed VCS?

A distributed VCS can have many topologies due to its distributed design (Figure 6.7). Any repository may be considered "the" repository since each is a peer to the others.

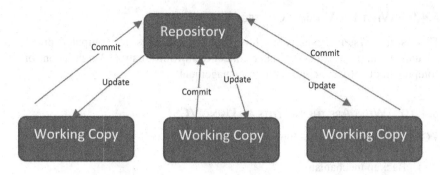

FIGURE 6.6   Centralized version control

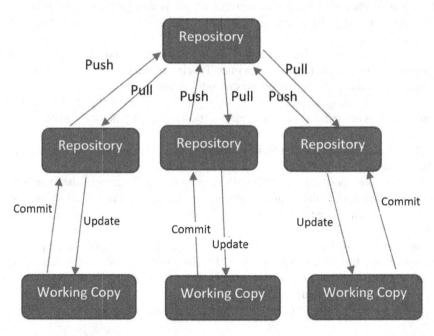

FIGURE 6.7   Distributed version control

The main feature of a distributed VCS is the support of workflows between different project members. This comes at the cost of having a two-level commit and update process.

6.4.13   ARE THERE OPEN-SOURCE, FREE VCSs?

Some open-source version control systems are available for free. These include SubVersion and CVS.

6.4.14   What Is Log Management Software?

Log Management Software is a tool that deals with a large volume of computer-generated messages. It is also known as event logs, audit trails, and audit records. Modern loggers have a wide range of functionality including:

- Support for different logging levels
- Logging to multiple data sinks
- Logging with timestamps and different formats
- Selective logging
- Automatic maintenance of log files.

6.4.15   Can You Give Examples of Log Management Tools?

- Site24x7: Supporting Windows, Linux, Mac
- Firewall Analyzer: Supporting Windows and Linux
- Datadog: Supporting Windows, Linux, Mac
- SolarWinds Log Analyzer: Supporting Windows
- Paessler Log Monitor: Supporting Linux, Mac, or Windows.

6.4.16   What Is a Build Tool?

A build tool is used to construct a software deliverable from its source files. An early build tool was **make** which would build C programs from its header and source files. This may sound easy, but recall that a C program requires:

- Preprocessing
- Compiling
- Linking.

The **make** tool also attempted to optimize its build by only compiling the modules of the system that had changed by examining the timestamps of files. Many modern build tools still use this technique.

Further, the **make** tool has this defining characteristic of a build tool—it runs unattended from the command line and produces a log indicating the success or failure of a build.

An integrated development environment aids a developer in writing code and compiling it properly. We usually don't consider an IDE as a build tool because it has a GUI interface and it doesn't scale well for very large projects.

6.4.17   What Are the Requirements to Look at When Selecting a Build Tool?

Here are some high-level requirements for a build tool. It must be:

- Scriptable: A readable script must drive a build tool. The script need not be a programming language. It can be a declarative script that the build tool understands and obeys.

- Repeatable: Once the build script has been finalized, it can be run repeatedly to produce the exact same outputs based on the same inputs. This may sound simple, but it is a common problem when one developer can build on one machine, but another developer finds that the same source will not build identical results on another machine.
- Scalable to Large Systems: This is where an IDE-based build typically fails. A large system may consist of 10M Lines of Code that constitute dozens of interdependent components. Each component must be built in the correct order using the correct dependencies.
- Able to manage dependencies between components: In the era of open-source software, it is common to use open-source components (e.g. jar files) to build commercial software. You want the commercial deliverable to be stable and that means you must know the version of every component that was used in the build.

Let's look at each of these.

6.4.18   ARE THERE OTHER TOOLS THAT I CAN USE, PARTICULARLY WHEN DEBUGGING EMBEDDED SYSTEMS?

A number of hardware and software tools are available to assist in the validation of embedded systems. Test tools make the difference between success and failure, especially in deeply embedded systems. For example, a multimeter can be used in the debugging of real-time systems to validate the analog input to or output from the system.

Oscilloscopes can be used for validating interrupt integrity, discrete signal issuance, and receipt, and for monitoring clocks. The more sophisticated storage oscilloscopes with multiple inputs can often be used in lieu of logic analyzers by using the inputs to track the data and address buses and synchronization with an appropriate clock.

Logic analyzers can be used to capture data or events, to measure individual instruction times, or to time sections of code. Programmable logic analyzers with integrated debugging environments further enhance the capabilities of the system integrator. Using the logic analyzer, the software engineer can capture specific memory locations and data for timing or for verifying the execution of a specific segment of code.

More sophisticated logic analyzers include built-in dissemblers and compilers for source-level debugging and performance analysis. These integrated environments are typically found on more expensive models, but they make the identification of performance bottlenecks particularly easy.

6.4.19   WHAT ARE IN-CIRCUIT EMULATORS?

During module-level debugging and system integration of embedded systems, the ability to single-step the computer, set the program counter, and insert into and read from memory is extremely important. This capability, in conjunction with the symbolic debugger, is the key to the proper integration of embedded systems. In an embedded environment, however, an in-circuit emulator (ICE) provides this capability.

An ICE uses special hardware in conjunction with software to emulate the target CPU while providing the aforementioned features. Typically, the ICE plugs into the chip carrier or card slot normally occupied by the CPU. External wires connect to an emulation system. Access to the emulator is provided directly or via a secondary computer.

6.4.20   How Are ICEs Used?

ICEs are useful in software patching and for single-stepping through critical portions of code. ICEs are not typically useful in timing tests, however, because the emulator can introduce subtle timing changes.

6.4.21   What Are Software Simulators and When Are They Used?

When integrating and debugging embedded systems, software simulators are often needed to stand in for hardware or inputs that do not exist or that are not readily available; for example, to generate simulated accelerometer or gyroscope readings where real ones are unavailable at the time.

The author of the simulator code has a task that is by no means easy. The software must be written to mimic exactly the hardware specification, especially in timing characteristics. The simulator must be rigorously tested; unfortunately, this is sometimes not the case. Many systems have been successfully validated and integrated with software simulators, only to fail when connected to the actual hardware.

6.4.22   When Is Hardware Prototyping Useful?

In the absence of the actual hardware system under control, simulation hardware may be preferable to software simulators. These devices might be required when the software is ready before the prototype hardware, or when it would be impossible to test the software on the actual hardware, such as in the control of a large nuclear plant.

Hardware simulators simulate real-life system inputs and can be useful for integration and testing but are not always reliable when testing the underlying algorithms; real data from live devices are needed.

6.4.23   What Are Integrated Development Environments?

Integrated development environments (IDEs) tie together various tools of the software production process through an easy-to-use GUI. IDEs can incorporate text editors, compilers, debuggers, coding standard enforcement, and much more. The most popular IDE includes the open-source Eclipse.

6.4.24   What about Other Tools?

There are many other commercial and open-source tools providing capabilities such as integrated document and software configuration control, testing management, stakeholder notification, distributed software built between cooperating PCs, and more.

6.5   OPEN SOURCE

6.5.1   WHAT IS OPEN-SOURCE SOFTWARE?

OSS is software that is free use or free redistribution if the terms of the license agreement are followed. Usually, this means that any work derived from the OSS can be redistributed only along with the source code and any derived works must comply with the same license.

OSS is the "opposite" of closed source software; that is, proprietary software for which the source cannot be had without paying a fee or had at all.

6.5.2   WHAT IS AN OPEN-SOURCE SOFTWARE LICENSE?

An open-source license sets forth the terms under which an individual or corporate entity can use a particular open-source software artifact. There are dozens of different kinds of open-source licenses—some grant the user unrestricted use of the software while others are quite restrictive. In some cases using open-source software with certain licenses requires the user to grant access to any derived artifacts or using the OSS code back to the repository.

Open-source license evaluation often requires legal counsel. Any user of open-source software needs to be very careful about how they use the software and contribute any software to the repository so as to protect intellectual property rights and comply with the licensing terms.

6.5.3   WHERE DID OSS COME FROM?

In 1983, Richard Stallman started the GNU project (which is a recursive acronym standing for GNU's Not Unix) to create a Unix-like operating system from free software. In 1991, Linus Torvalds created an operating system called Linux while a graduate student at the University of Helsinki. Along the way, the process and culture created by Stallman and others, and carried on by Torvalds, formed the basis for the OSS movement today.

6.5.4   WHAT KINDS OF CODE CAN BE FOUND AS OPEN SOURCE?

There are many thousands of open-source projects ranging from games to programming languages, tools, and enterprise-level applications. Clones of many well-known desktop and enterprise applications are also available in open-source. The different types of OSS projects are summarized in Table 6.3.

- Linux operating system
- Firefox Web browser
- Apache Web server
- GCC, GNU C compiler.

For the software engineer, there are many scripting languages like Perl, Python, PHP, and Ruby available in open-source. Other useful developer tools include Ant and

TABLE 6.3
Various Kinds of OSS Projects

Type	Objective	Control Style	Community Structure	Major Problems	Examples
Exploration-oriented	Sharing innovation and knowledge	Cathedral-like Central control	Project leader Many readers	Subject to split	GNU systems JUN Perl
Utility-oriented	Satisfying an individual need	Bazaar-like Decentralized control	Many peripheral developers Peer support to passive users	Difficult to choose the right program	Linux system (excluding the kernel = exploration-oriented)
Service-oriented	Providing stable services	Council-like Central control	Core members instead of a project leader Many passive users that develop systems for end users	Less innovation	Apache PostgreSQL

Source: Nakakoji, Yamamoto Nishinaka, Kishida, and Ye, 2002.

Maven for building applications; Hibernate, which acts as an object-oriented persistence layer for databases; Xunit for testing; CVS and SubVersion for source code control; and Eclipse or NetBeans as integrated development environments.

The software engineer may also want to use the open-source Struts, which provides a framework in which an application can be built quickly using the model-view-controller architecture, and Swing, which provides a layered structure for managing business objects.

6.5.5 WHAT IS THE VALUE PROPOSITION FOR OSS?

The benefits of using OSS are clear access to a large amount of sophisticated and useful code for free or nearly free and, for those applications with the most robust communities, it can be expected that the code will be maintained for many years.

OSS advocates also claim that OSS has fewer defects than closed source code. They quote Eric Raymond (2001) in this regard: "given enough eyeballs, all bugs are shallow." Of course, it is not true that OSS is error-free; some is quite buggy. But the research is still wide open on whether open-source code is uniformly better, or worse than closed source code. The truth is, it probably depends on the project.

6.5.6 WHO CONTRIBUTES TO OSS SYSTEMS?

Many people contribute code to open-source repositories (or act as testers and beta users) for many different reasons. Individuals get involved to do something "important" or to be part of a community. Companies allow their employees to contribute code because the code has some benefit to the company. Others participate simply for fun or self-interest.

6.5.7 Where Can I Find OSS Projects?

OSS projects can be found in any number of open-source repositories. Github is the largest, most popular code repository on the web. SourceForge is also a huge code repository, particularly suitable for open-source projects thanks to supporting for Git, Mercurial, and SubVersion versioning systems.

6.5.8 Can Companies Use OSS?

Most companies have some OSS in play, even if they decry its use. Many "open-source forbidden" enterprises run Apache, use some of the open languages, or take advantage of open-source Java libraries.

6.5.9 What Are the Characteristics of the OSS Development Model?

Characterization of the OSS development model is based on a model proposed by Eric Raymond (2001). Raymond contrasts the traditional software development of a few people planning a cathedral in splendid isolation with the new collaborative "bazaar" form of OSS development. OSS systems co-evolve with their developer communities, though the usual team critical mass of 5 to 15 core developers is needed to sustain a viable system.

6.5.10 How Does Software Requirements Engineering Occur in OSS?

Software requirements for OSS usually take the form of threaded messages and Web site discussions. Requirements engineering is closely tied to the interests of the OSS developer community (Scacchi, 2004).

6.5.11 How Do the Software Design and Build Processes Take Place in Open-Source Systems?

OSS code contributors (called committers) are often end users of the software. These committers usually employ versions of open-source development tools such as CVS to empower a process of system build and incremental release review. Other members of the community use open-source bug reporting tools (e.g., Bugzilla) to dynamically debug the code and propose new features. Concurrent versions of systems play an essential role in the coordination of decentralized code.

Software features are added to evolutionary redevelopment, reinvention, and revitalization. The community of developers for each open-source project generally experiences a high degree of code sharing, review, modification, and redistributing concepts and techniques through minor improvements and mutations across many releases with short life cycles (Scacchi, 2004).

6.5.12 How Are OSS Projects Managed?

OSS project management consists of an organized interlinked layered meritocracy. This configuration is a hierarchical organizational form that centralizes and concentrates certain kinds of authority, trust, and respect for experience and accomplishment within the team. Such an organization is a highly adaptive but loosely coupled virtual enterprise (Scacchi, 2004).

6.5.13 Are There Downsides to Using OSS?

Because of the dynamic nature of OSS, code versions and patches need to be carefully managed for any software that is adopted in the enterprise. There are companies, however, that manage these artifacts and provide "certified solution stacks" (collections of interoperable OSS) for a fee. Unless you contract such a company, however, the maintenance and security issues associated with using OSS applications could become a nightmare.

As far as using OSS in an end product, the licensing issues need to be well understood and managed. If they are not well understood, then the company may be at risk for legal troubles.

There is one other downside. Aside from the major OSS applications, which are highly sophisticated and developed, many of the software projects in open-source repositories contain many bugs or lack important features because they are in various stages of maturity with developers of varying skill and motivation levels.

Despite the downsides, OSS or a blend of open-source and new should be considered as part of any new software development initiative.

6.6 BECOMING A BETTER CODE DEVELOPER

6.6.1 How Can I Become a Better Developer of Code?

Coding software is not software engineering. The best code starts out with a good design. That being said, it is always important to improve your mastery of the programming language, just as you can improve your ability to write in English by better understanding the rules of the language, improving your vocabulary, and reading great works in the language. The best way to improve coding skills is to practice and read the appropriate literature. That raises one other point: one of the best ways to learn a programming language is to start by reading great code that someone else has written.

6.6.2 Code Smells

6.6.2.1 What Is a Code Smell?

A code smell refers to an indicator of poor design or coding (Fowler, 2000). More specifically, the term relates to visible signs that suggest the need for refactoring. We will discuss refactoring in the next subsection. Code smells are found in every kind of system.

6.6.2.2 What Are Some Code Smells?

Table 6.4 summarizes a set of code smells originally described for object-oriented languages by Fowler (2000). But many of these code smells are also found in procedural languages.

A few of these and several others not identified by Fowler that are found in procedural systems and described by Stewart (1999) will be discussed in terms of C, which

---

TABLE 6.4
Some Code Smells and Their Indicators

Code Smell	Description
Alternative classes with different interfaces	Methods that do the same thing but have different signatures
Conditional logic	Switch and case statements that breed duplication
Data class	Classes with just fields, getters, setters, and nothing else
Data clumps	Several data items found together in lots of places; should be an object
Divergent change	When a class is changed in different ways for different reasons
	Should be that each object is changed only for one type of change
Duplicated code	The same code in more than one place
Feature envy	Objects exit to package data with the processes used on that data
Inappropriate intimacy	When classes spend too much time interacting with each others' private parts
Incomplete library class	Poorly constructed library classes that are not reused
Large class	A class that is doing too much
Lazy class	A class that isn't doing enough to "pay" for itself
Long method	Short methods provide indirection, explanation, sharing, and choosing
Long parameter list	Remnant of practice of using parameter lists vs. global variables
Message chains	The client is coupled to the structure of the navigation: `getA().getB().getC().getD().getE().doIt()`
Middle man	When a class delegates too many methods to another class
Parallel inheritance hierarchies	Special case of shotgun surgery
	Need to ensure that instances of one hierarchy refer to instances of the other
Primitive obsession	Aversion to using small objects for small tasks
Refused bequest	When a child only wants part of what is coming from the parent
Shotgun surgery	Every time you make a change you have to make a lot of little changes
	Opposite of divergent change
Speculative generality	Hooks and special cases to handle things that are not required (might be needed someday)
Tell-tale comments	Comments that exist to hide/explicate bad code
Temporary field	Instance variables that are only set sometimes; you expect an object to use all of its variables

Source: Fowler, 2000.

---

has many constructs in common with C++, Java, and C#, so most readers should be able to follow the code fragments.

6.6.2.3   What Is the Conditional Logic Code Smell?

These are excessive `switch`, `if-then`, and `case` statements and they are an indicator of bad design for several reasons. First, they breed code duplication. Moreover, the code generated for a case statement can be quite a convoluted—for example, a jump through a register, offset by a table value. This mechanism can be time-consuming. Furthermore, nested conditional logic can be difficult to test, especially if it is nested due to a large number of logic paths through the code. Finally, the differences between best and worst-case execution times can be significant, leading to highly pessimistic utilization figures.

Conditional logic needs to be refactored, but there is no silver bullet here. In object-oriented languages, the mechanisms of polymorphism or composition can be used. In procedural languages, sometimes mathematical identities can be used to improve the efficiency of the code.

6.6.2.4   What Are Data Clumps?

Several data items found together in lots of places are known as data clumps. For example, in the procedural sense, data clumps can arise in C from too much configuration information in `#include  files`. Stewart (1999) notes that this situation is unhealthy because it leads to increased development and maintenance time and introduces circular dependencies that make reuse difficult. He suggests that to refactor, each module be defined by two files, `.h` and `.c`, with the former containing only the information that is to be exported by the module and the latter containing everything that is not exported.

A second manifestation of the data clump smell has to do with excessive use of the `#define` statement that propagates throughout the code. Suppose these `#defines` are expanded in 20 places in the code. If, during debugging, it is desired to place a patch over one of the `#defines`, it must be done in 20 places. To refactor, place the quantity in a global variable that can be changed in one place only during debugging.

6.6.2.5   Why Are Delays as Loops Bad?

Stewart (1999) suggests another code smell involving timing delays implemented as `while` loops with zero or more instructions. These delays rely on the overhead cost of the loop construct plus the execution time of the body to achieve a delay. The problem is that if the underlying architecture or characteristics of instruction execution (e.g., memory access time change) changes, then the delay time is inadvertently altered. To refactor, use a mechanism based on a timer facility provided by the operating system that is not based on individual instruction execution times.

6.6.2.6   What Are Dubious Constraints?

This code smell is particularly insidious in embedded systems where response time constraints have a questionable or no attributable source. In some cases, systems have deadlines that are imposed on them that are based on nothing more than guessing or on some forgotten and since eliminated requirement. The problem in these cases is that undue constraints may be placed on the systems.

For example, suppose the response time for an event is 30 ms but no one knows why. Similarly, more than one reason given for the constraints in comments or documentation indicates a traceability conflict, which hints at other problems. This is a primary lesson in embedded systems design to understand the basis and nature of the timing constraints so that they can be relaxed if necessary.

It is typical, in building embedded systems, to consider the nature of time because deadlines are instants in time. But the question arises, "where do the deadlines come from?" Generally speaking, deadlines are based on the underlying physical phenomena of the system under control. For example, in animated displays, real-time images must be updated at approximately 30 frames per second to provide a continuous image because the human eye can resolve to update at a slower rate. In navigation systems, accelerations must be read at a rate that is based on the top speed of the vehicle, and so on.

In any case, to remove the code smell, some detective work is needed to discover the true reason for the constraint. If it cannot be determined, then the constraint could be relaxed and the system redesigned accordingly.

6.6.2.7   What Is the Duplicated Code Smell?

Obviously, duplicated code refers to the same or similar code found in more than one place. It has an unhealthy impact on maintainability (the same change has to be propagated to each copy), and it also adversely affects memory utilization. To refactor, assign the code to a single common code unit via better application of information hiding.

While it is too easy to mock the designers of systems that contain duplicated code, it is possible that the situation arose out of a real need at the time. For example, duplicated code may have been due to legacy concerns for performance where the cost of the procedure call added too much overhead in a critical instance. Alternatively, in an older version of languages such as Fortran that were not reentrant, duplicated code was a common means for providing utilities to each cycle in the embedded system.

6.6.2.8   What Are Generalizations Based on a Single Architecture?

Stewart (1999) suggests that writing software for a specific architecture, but with the intent to support easy porting to other architectures later, can lead to over-generalizing items that are similar across architectures, while not generalizing some items that are different. He suggests developing the code simultaneously on multiple architectures and then generalizing only those parts that are different. He also suggests choosing three or four architectures that are very different in order to obtain the best generalization. Presumably, such an approach suggests the appropriate refactoring.

6.6.2.9   What Are the Large Method, Large Class, and Large Procedure Code Smells?

Fowler (2000) describes two code smells, long method and large class, which are self-evident. In the procedural sense, the analogy is a large procedure. Large procedures are anathema to the divide and conquer principle of software engineering and need to be refactored by re-partitioning the code appropriately.

6.6.2.10   What Are Lazy Methods and Lazy Procedures?

A lazy method is one that does not do enough to justify its existence. The procedural analogy to the lazy class/method code smells is the lazy procedure. In a real-time sense, a procedure that does too little to pay for the overhead of calling the procedure needs to be eliminated by removing its code to the calling procedure or redefining the procedure to do more.

6.6.2.11   What Is the Long Parameter List Code Smell and How Can It Be Refactored?

Long parameter lists are an unwanted remnant of the practice of using parameter lists to avoid the use of global variables. While clearly well-defined interfaces are desirable, long parameter lists can cause problems in embedded systems if interrupts are disabled during parameter passing. In this case, overly long interrupt latencies and missed deadlines are possible.

The long parameter list code smell can be refactored by passing a pointer to one or more data structures that contain aggregated parameters, or by using global variables.

6.6.2.12   What Are Message Chains?

Fowler (2000) describes message chains as a bad smell in OOD. For example, a message chain occurs if the client is coupled to the structure of the navigation: getA().getB().getC().getD().getE().doIt();.

The procedural analogy to message chains might be an overly long sequence of procedure calls that could be short-circuited or replaced by a more reasonable sequence of calls. The problem in the case of long chains is that the overhead of calling procedures becomes significant, interrupts may be disabled during parts of the procedure calls (e.g., for parameter passing), and the long sequence of calls may indicate inefficient design.

6.6.2.13   What Is Message-Passing Overload?

Stewart (1999) describes the excessive use of message-passing for synchronization as another unwanted practice. He notes that this practice can lead to unpredictability (because of the necessary synchronization), the potential for deadlock, and the overhead involved. He suggests that the refactoring is to use state-based communication via shared memory with structured communication.

6.6.2.14   What Is the One Big Loop Code Smell and How Is It Refactored?

Cyclic executives are non-interrupt-driven systems that can provide the illusion of simultaneity by taking advantage of relatively short processes on a fast processor in a continuous loop. For example, consider the set of self-contained processes Process1 through Processn in a continuous loop as depicted in the following:

```
for(;;)  {    /* do forever */
         Process1();
         Process2();

         ...Processn()
         }

}
```

Stewart (1999), who calls the cyclic executive "one big loop," notes that there is little flexibility in the cyclic executive and that only one cycle rate is created. However, different rate structures can be achieved by repeating a task in the list. For example, in the following code

```
for(;;)  {    /* do forever */
     Process1();
     Process2();
     Process3();
     Process3();
             }
      }
```

Process3 runs twice as frequently as Process1 or Process2.

Moreover, the task list can be made flexible by keeping a list of pointers to processes, which can be managed by the "operating system" as tasks are created and completed. Inter-task synchronization and communication can be achieved through global variables or a parameter list.

Generally, however, the "big loop" structure really is a bad smell unless each process is relatively short and uniform in size. The refactoring involves rebuilding the executive using an interrupt scheme such as rate-monotonic or round-robin.

6.6.2.15   What Is Shotgun Surgery?

Shotgun surgery is a very common code smell related to the phenomenon that every time you make a change you have to make many little changes. This is another example of poor application of information hiding, which suggests refactoring.

6.6.2.16   What Is Speculative Generality?

Speculative generality relates to hooks and special cases that are built into the code to handle things that are not required (but might be needed someday).

Embedded systems are no place to build in hooks for "what-if" code. Hooks lead to testing anomalies and possible unreachable code. Therefore, the refactoring is to remove hooks and special cases that are not immediately needed.

6.6.2.17   What Are Tell-Tale Comments?

The tell-tale comment problem appears in all kinds of systems. Comments that are excessive, or which tend to explicate the code beyond a reasonable level, are often indicators of some serious problem. Comments such as "do not remove this code," or "if you remove this statement, the code doesn't work, I don't know why" are not uncommon. Humor in comment statements can sometimes be a glib way to mask the fact that the writer doesn't know what he is doing. Oftentimes these kinds of statements indicate that there are underlying timing errors.

In either case, the refactoring involves rewriting the code so that an overly long explicating comment is not necessary.

6.6.2.18 What Are Temporary Fields and How Are They Refactored?

Temporary fields are instance variables that are only set sometimes; you expect an object to use all of its variables. In the procedural sense, this could be seen in the case of a `struct` with fields that are unused in certain instances. The refactoring is to use an alternative data structure. For example, in C you could use a `union` to replace the struct.

6.6.2.19 What Is the Unnecessary Use of Interrupts Code Smell?

Stewart (1999) also suggests that indiscriminate use of interrupts is a bad code smell. Interrupts can lead to deadlock, priority inversion, and inability to make performance guarantees (Laplante, 2004b).

Interrupt-based systems should be avoided, although it has been noted that avoidance is possible only in the simplest of systems where real-time multi-tasking can be achieved with coroutines and cyclic executives implemented without interrupts. When interrupts are required to meet performance constraints, rate-monotonic or earliest deadline first scheduling should be used.

6.6.2.20 How Can I Improve the Runtime Performance of the Code I Write?

Many of these improvements can be achieved by refactoring the code smells just discussed. However, the main thing you can do to improve code execution performance, particularly in embedded systems, is to understand the mapping between high-order language input and assembly language output for a particular compiler. This understanding is essential in generating code that is optimal in either execution time or memory requirements. The easiest and most reliable way to learn about any compiler is to run a series of tests on specific language constructs.

For example, in many C compilers the `case` statement is efficient only if more than three cases are to be compared; otherwise, nested `if` statements should be used. Sometimes the code generated for a `case` statement can be quite convoluted; for example, a jump through a register offset by the table value. This sequence can be time-consuming.

It has already been noted that procedure calls are costly in terms of the passing of parameters via the stack. The software engineer should determine whether the compiler passes the parameters by byte or by word.

Other language constructs that may need to be considered include:

- Use of while loops vs. for loops or do-while loops.
- When to "unroll" loops; that is, to replace the looping construct with repetitive code (thus saving the loop overhead as well as providing the compiler the opportunity to use faster machine instructions).

- Comparison of variable types and their uses (e.g., when to use short integer in C vs. Boolean, when to use single precision vs. double precision floating point, and so forth).
- Use of in-line expansion of code via macros vs. procedure calls. This is by no means an exhaustive list of tips.

6.6.3 REFACTORING

6.6.3.1 What Is Refactoring?

The term "refactoring" applies to many situations. It has been described as:

- The process of changing a software system in such a way that it does not alter the external behavior of the code, yet it improves the internal structure (Fowler, 2018).
- A behavior preserving source-to-source program transformation (Roberts, 1999).
- A change to the system that leaves its behavior unchanged, but enhances some non-functional quality—simplicity, flexibility, understandability, … (Beck & Andres, 2004).

So we see that it is both a noun and a verb. It is the "behavior preserving transformation" and it is the process of transformation itself.

6.6.3.2 What Does the Process of Refactoring Involve?

The process of refactoring involves the removal of duplication, the simplification of complex logic, and the clarification of unclear code. When you refactor, you relentlessly, as Kent Beck notes "poke and prod your code to improve its design" (Beck & Andres, 2004). Such improvements may involve something as small as changing a variable name or as large as unifying two hierarchies.

To refactor safely, you must either manually test that your changes didn't break anything or run automated tests. You will have more courage to refactor and be more willing to try experimental designs if you can quickly run automated tests to confirm that your code still works. Martin Fowler (2018) classifies three levels for refactoring: Low level, Medium level, and High level.

6.6.3.3 What Is Low-Level Refactoring?

The low-level refactoring reflects good, initial coding practices. Some of the refactoring activities at this level include:

- Replace a "magic[1]" number with a named constant (or enumerated values)
- Rename a variable with a clearer or more informative name
- Convert a set of codes to a full-featured enumeration
- Move complex Boolean expressions into a well-named Boolean function.

6.6.3.4 What Is Medium-Level Refactoring?

Some of the activities involved in this level include:

- Consolidate duplicate code into a common method
- Decompose a long method into multiple shorter methods
- Combine similar routines by parameterizing them
- Remove set routines for fields that cannot change.

6.6.3.5 What Is High-Level Refactoring?

Refactoring at this level involves object-oriented design principles and patterns. Activities at this level include:

- Decompose a Large Class into Smaller Classes.
- Extract superclass: At times you may discover an inheritance hierarchy while programming. If this is the case, then you can "factor" out the code that belongs to the superclass. This sort of refactoring is commonly aided by the IDE that you are using.
- Extract interface: This refactoring has legitimate uses at a time; however, sometimes it is used as a trick by programmers to break circular dependencies between two classes.

6.6.3.6 Is Refactoring Limited to the Code and Low-Level Design?

The realization that (tactical) refactoring, such as that described by Fowler, is ineffective against large-scale design decay gave rise to the concept of strategic refactoring. This focuses on architectural rejuvenation through refactoring. The steps to strategic refactoring are as follows:

1. Analyze existing application for "bad smells"
2. Determine target 'gross' architecture (Layers, MVC, etc.)
3. Define target 'macro' architecture (domain layer roles, etc.)
4. Identify micro-architectural patterns to implement changes
5. Tease away.

By thinking strategically we can apply collections of low-level transforms to decaying systems and revive them. The process is iterative, where the first iteration is usually concerned with reducing opacity. Once the system is comprehensible, we can address other properties such as reuse, performance, extension, security, and dependability.

6.6.4 CODING STANDARDS

6.6.4.1 What Are Coding Standards?

Coding standards are different from language standards. A language standard (e.g., ANSI C*) embodies the syntactic rules of the language. A program violating those rules will be rejected by the compiler. Conversely, a coding standard is a set of stylistic conventions. Violating the conventions will not lead to compiler rejection. In another sense, compliance with language standards is mandatory while compliance with coding standards is voluntary.

6.6.4.2   How Can Coding Standards Help Improve My Code?

Adhering to language standards fosters portability across different compilers and, hence, hardware environments. Complying with coding standards will not likely increase portability, but rather in many cases will increase readability and maintainability. Some even contend that the use of coding standards can enhance reliability. Coding standards may also be used to promote improved performance by encouraging or mandating the use of language constructs that are known to generate code that is more efficient. Many agile methodologies, for example, Extreme Programming, embrace coding standards.

6.6.4.3   What Do Coding Standards Look Like?

Coding standards involve standardizing some or all of the following elements of programming language use:

- Standard or boilerplate header format
- Frequency, length, and style of comments
- The naming of classes, methods, procedures, variable names, data, file names, and so forth
- Formatting of program source code including use of white space and indentation
- Size limitations on code units include maximum and minimum lines of code, number of methods, and so forth.
- Rules about the choice of language construct to be used; for example, when to use case statements instead of nested if-then-else statements

This is just a partial list.

6.6.4.4   What Is the Benefit of Coding Standards?

Close adherence to a coding standard can make programs easier to read and understand and likely more reusable and maintainable.

6.6.4.5   Which Coding Standard Should I Use?

Many different standards for coding are language-independent or language-specific. Coding standards can be team-wide, company-wide, or user group specific. For example, the GNU software group has standards for C and C++, or customers can also require conformance to a specific standard that they own. Still, other standards have become public domain.

One example is the Hungarian notation standard, named in honor of Charles Simonyi who is credited with first promulgating its use. Hungarian notation is a public domain standard intended to be used with object-oriented languages, particularly C++. The standard uses a complex naming scheme to embed type information about the objects, methods, attributes, and variables in the name. Because the standard essentially provides a set of rules about naming variables, it can be used with other languages such as C++, Ada, Java, and even C.

6.6.4.6 Are There Any Drawbacks to Using Coding Standards?

First, let me say that I believe that you should always follow a coding standard, despite any difficulties they might introduce or shortcomings they might have. One problem with standards, however, is that they can promote very mangled variable names. In other words, the desire to conform to the standard is greater than creating a particularly meaningful variable name. Another problem is that the very strength of coding standards can be their undoing. For example, in Hungarian notation, what if the type information embedded in the object name is wrong? There is no way for a compiler to check this. There are commercial rule wizards that can be tuned to enforce the coding standards, but they must be tuned to work in conjunction with the compiler.

6.6.4.7 When Should the Coding Standard Be Adopted?

Adoption of coding standards is not recommended mid-project. It is much easier to conform at the start of the project than to be required to change existing code.

NOTE

1   Some important constant, such as e or, algebraic combination of constants such as 2π.

FURTHER READING

Beck, K. & Andres, C. (2004). *Extreme Programming Explained: Embrace Change*. 2nd edition. Addison-Wesley Professional.

Cardelli, L. (1996). Bad engineering properties of object-oriented languages. *ACM Comp. Surveys, 28A*(4), 150–158.

Chen, Y., Dios, R., Mili, A., Wu, L., & Wang, K. (2006). An empirical study of programming language trends. *IEEE Software, 22*(3), 72–78.

Fowler, M. (2000). *Refactoring*. Addison-Wesley, Boston, MA.

Fowler, M. (2018). *Refactoring: improving the design of existing code*. Addison-Wesley Professional.

Gamma, E., Helm, R., Johnson, R., & Vlissides, J. (1995). *Design Patterns: Elements of Reusable Object-Oriented Software*. Addison-Wesley, Boston, MA.

Kernighan, B.W. & Ritchie, D.M. (1988). *The C Programming Language*. 2nd ed., Prentice Hall, Englewood Cliffs, NJ.

Laplante, P.A. (2004a). *Software Engineering for Image Processing System*. CRC Press, Boca Raton, FL.

Laplante, P.A. (2004b). *Real-Time Systems Design and Analysis: An Engineer's Handbook*. 3rd ed., IEEE Press/John Wiley & Sons, New York.

Louden, K.C. & Lambert, K. (2011). *Programming Languages: Principles and Practice*. 3rd ed., Cengage Learning, Boston, MA.

McConnell, S. (2004). *Code Complete*. 2nd ed., Microsoft Press, Redmond, WA.

Meyer, B. (2000). *Object-Oriented Software Construction*. 2nd ed., Prentice Hall, Englewood Cliffs, NJ.

Nakakoji, K., Yamamoto, Y., Nishinaka, Y., Kishida, K., & Ye, Y. (May 2002). Evolution patterns of open-source software systems and communities. *Proc. Intl. Workshop on Principles of Software Evolution*, ACM Press, Orlando, FL.

Roberts, D.B. (1999). *Practical analysis for refactoring.* University of Illinois at Urbana-Champaign.

Scacchi, W. (2004). Understanding open-source software evolution. In *Software Evolution*. Madhavji, N.H., Lehman, M.M., Ramil, J.F., & Perry, D. (Eds.), John Wiley & Sons, New York.

Sebesta, R.W. (2015). *Concepts of Programming Langua^{ge}.*, 11th ed., Pearson, London, UK.

Stewart, D.B. (1999). Twenty-five most common mistakes with real-time software development. *Class #270, Proc.* 1999 Embedded Syst. Conf., San Jose, CA.

Tate, B. (2010). *Seven languages in seven weeks: a pragmatic guide to learning programming languages.* Pragmatic Bookshelf.

7 Software Quality Assurance

OUTLINE

- Quality models and standards
- Software testing
- Metrics
- Fault tolerance
- Maintenance and reusability

7.1 INTRODUCTION

"In the beginning of a malady it is easy to cure but difficult to detect, but in the course of time, not having been either detected or treated in the beginning, it becomes easy to detect but difficult to cure" (Machiavelli, 1513). Machiavelli's sentiments are precisely those that must abound in a software enterprise seeking to produce a quality product. Apparently, though, this sentiment is not prevalent. A 2002 study by the National Institute of Standards Technology (NIST) estimated that software errors cost the U.S. economy $59.5 billion each year. The report noted that software testing could have reduced those costs to about $22.5 billion. Of the $59.5 billion, users paid for 64%, and developers paid for 36% of the cost (NIST, 2002).

Software failures due to lack of testing have wreaked havoc in almost every domain of business with cases including even loss of people's lives as in the recent events involving Boeing 737 Max airplanes. A detailed review investigating a representative set of 59 recent catastrophic accidents due to undiscovered software bugs was recently presented in Wong, Li, & Laplante (2017) along with lessons learned and implications for future software systems.

In order to achieve the highest levels of software quality, there are several things you have to do. First, you need to have in place a system that will foster the development of quality software. This is what the quality models are all about. Incorporated in that quality software system is rigorous, life cycle testing.

In order to attain any of the desirable software qualities, you must have a way to measure them through metrics. Finally, you can improve the reliability of the software through a fault-tolerant design. This chapter discusses all of these aspects of software quality.

DOI: 10.1201/9781003218647-8

7.2　QUALITY MODELS AND STANDARDS

7.2.1　WHAT IS SOFTWARE QUALITY?

There are many ways that stakeholders might perceive software quality, all based on the presence or absence to a certain degree of one attribute or another. Some formal definitions are appropriate, however, as defined in the *ISO 8402:2000 Quality Management and Quality Assurance — Vocabulary standard*.

We have already discussed software quality in chapters 2 and 3. A quality policy describes the overall intentions and direction of an organization with respect to quality, as formally expressed by top management. Quality management is that aspect of the overall management function that determines and implements quality policy. Finally, a quality system comprises the organizational structure, responsibilities, procedures, processes, and resources for implementing quality management.

As it turns out, one can undermine any of these definitions with rhetorical arguments but, for our purposes, they are useful working definitions. These are broad definitions of quality for every kind of product, not just software. There are other ways to look at software quality.

For example, Voas and Agresti (2004) propose that quality comprises a set of key behavioral attributes such as:

- Reliability (R)
- Performance (P)
- Fault tolerance (F)
- Safety (Sa)
- Security (Se)
- Availability (A)
- Testability (T)
- Maintainability (M).

along with a dash of uncertainty. They further suggest that quality has a slightly different meaning for each organization, and perhaps each application, which can be represented by a weighted linear combination of these behavioral attributes, namely,

$$Q = w_1R + w_2P + w_3F + w_4Sa + w_5Se + w_6A + w_7T + w_8M + \text{uncertainty}.$$

Munson (2003) discusses quality as a set of objectives:

- Learn to measure accurately people, processes, products, and environments
- Learn to do the necessary science to reveal how these domains interact
- Institutionalize the process of learning from past mistakes
- Institutionalize the process of learning from past successes
- Institutionalize the measurement and improvement process.

All of these different takes on quality are compatible and largely consistent with the prevailing quality models.

7.2.2 What Is a Quality Model?

A quality model is a system for describing the principles and practices underlying software process maturity. A quality model is different from a life cycle model (we discussed that in Chapter 2) in that the latter is used to describe the evolution of code from conception through delivery and maintenance. A quality model is intended to help software organizations improve the maturity of their software processes as an evolutionary path from ad hoc, chaotic processes to mature, disciplined software processes.

Some famous and widely employed quality models in the software industry are:

- The capability maturity model (CMM)
- ISO 9000
- ISO/IEC 12207
- Six Sigma
- IT infrastructure library (ITIL).

7.2.3 What Is the Capability Maturity Model?

The capability maturity model CMM is a software quality model consisting of five levels. Predictability, effectiveness, and control of an organization's software processes are believed to improve as the organization moves up these five levels. While not truly rigorous, there is some empirical evidence that supports this position.

The CMM for software is not a life cycle model, but rather a system for describing the principles and practices underlying software process maturity. Like other quality models, CMM is intended to help software organizations improve the maturity of their software.

CMM had its foundations in work begun in 1986 at the U.S. DoD to help improve the quality of the deliverables produced by government software contractors. The work was commissioned through the MITRE Corporation but later moved to the Software Engineering Institute (SEI) at Carnegie Mellon University. Watts Humphrey was the initial author, and then Mark Paulk, Bill Curtis, and others took the lead in CMM's development (Paulk, Weber, Curtis & Chrissis, 1995). CMM borrows heavily from general Total Quality Management (TQM) and the work of Philip Crosby. Version 2.0 of CMM was published in 2018.

The five maturity levels in the model are:

- Initial
- Managed
- Defined
- Quantitatively managed
- Optimizing

Each maturity level is a well-defined evolutionary stage of a mature software process. The maturity levels indicate process capabilities, which describe the range of expected results that can be achieved by following a software process.

7.2.4 What Is ISO 9000?

ISO 9000 is a generic, worldwide standard for quality improvement and it is based on seven quality management principles (QMPs):

- QMP 1 – Customer focus
- QMP 2 – Leadership
- QMP 3 – Engagement of people
- QMP 4 – Process approach
- QMP 5 – Improvement
- QMP 6 – Evidence-based decision making
- QMP 7 – Relationship management.

The International Standards Organization owns the standard. Collectively described in five standards, ISO 9000 through ISO 9004, the ISO 9000 series was designed to be applied in a wide variety of manufacturing environments. While ISO 9000 is the set of standards that basically describes a quality management system, ISO 9001 specifies the requirements and criteria for the certification. Third-party certification bodies provide independent confirmation that organizations meet the requirements of ISO 9001. In 2014, there were approximately 1,138,200 organizations globally considered as ISO 9001 certified.

7.2.5 What Are the ISO 90003 Principal Areas of Quality Focus?

ISO/IEC/IEEE 90003:2018 is essentially an expanded version of ISO 9001 with added narrative to encompass software. They are:

- Management responsibility
- Quality system requirements
- Contract review requirements
- Product design requirements
- Document and data control
- Purchasing requirements
- Customer supplied products
- Product identification and traceability
- Process control requirements
- Inspection and testing
- Control of inspection, measuring, and test equipment
- Inspection and test status
- Control of nonconforming products
- Corrective and preventive actions
- Handling, storage, and delivery
- Control of quality records
- Internal quality audit requirements
- Training requirements
- Servicing requirements
- Statistical techniques.

ISO 9000-3	4.4 Software development and design
4.4.1 General	**Develop and document procedures to control the product design and development process. These procedures must ensure that all requirements are being met.**
Software development	Control your software development project and make sure that it is executed in a disciplined manner.

- Use one or more life cycle models to help organize your software development project.
- Develop and document your software development procedures. These procedures should ensure that:
  - Software products meet al requirements.
  - Software development follows your:
    - Quality plan.
    - Development plan.

FIGURE 7.1   Excerpt from ISO 90003: 4.4 Software development and design

7.2.6   What Does ISO 90003 Look Like?

It is short (approximately 30 pages) and very high level (see Figure 7.1 for an excerpt).

7.2.7   How Does ISO/IEC 12207 Help Promote Software Quality?

This standard, which was briefly introduced in Chapter 2, is contained in a relatively high-level document used to help organizations refine their processes, by defining compliance as the performance of those processes, activities, and tasks. Hence, it allows an organization to define and meet its own quality standards. Because it is so general, organizations seeking to apply 12207 need to use additional standards or procedures that specify those details.

7.2.8   What Is Six Sigma?

Developed by Motorola, Six Sigma is a management philosophy based on removing process variation. Six Sigma focuses on the control of a process to ensure that outputs are within six standard deviations (Six Sigma) from the mean of the specified goals. Six Sigma is implemented using define, measure, analyze, improve, and control (DMAIC).

Define means to describe the process to be improved, usually through some sort of business process model. *Measure* means to identify and capture relevant metrics for each aspect of the process model. The goal question metric paradigm is helpful in this regard. *Analyze* means to review the data. *Improve* means to change some aspect of the process so that beneficial changes are seen in the associated metrics, usually by attacking the aspect that will have the highest payback. Finally, *control* mean to use ongoing monitoring of the metrics to continuously revisit the model, observe the metrics, and refine the process as needed.

Some organizations use Six Sigma as part of their software quality practices. The issue here, however, is in finding an appropriate business process model for the software production process that does not devolve into a simple, and highly artificial, waterfall process.

7.2.9   WHAT IS THE IT INFRASTRUCTURE LIBRARY?

The IT Infrastructure Library (ITIL) is a worldwide standard for IT service management. Originating in the U.K., ITIL has standards for:

- Service support
- Service delivery
- Infrastructure management
- Application management
- Planning to implement
- Business perspective.

"Dashboard" tools are available for each of these standards.

7.2.10   CAN ITIL HELP WITH SOFTWARE QUALITY PROGRAMS?

Some companies find ITIL to be a useful framework for software quality management because it is non-proprietary, platform-independent, and adaptable.

7.2.11   HOW DOES ITIL HELP WITH SOFTWARE QUALITY MANAGEMENT?

ITIL helps create comprehensive, consistent, and coherent codes of best practice for quality IT service management, promoting business effectiveness in the use of IT. ITIL also encourages the private sector to develop ITIL-related services and products, training, consultancy, and tools.

7.2.12   CAN ANYTHING BAD COME OF A SOFTWARE QUALITY INITIATIVE?

Yes. According to West (2004), "slash-and-burn improvement" can occur. The slash-and-burn approach means throwing away the good practices (and perhaps people) along with the bad; for example, when it is determined (or assumed) that an organization is at CMM Level 1, and the decision is made to tear down the existing software culture and start from scratch. The negative effects of slash-and-burn process improvement are not always easy to detect (West, 2004).

7.2.13   WHAT ARE THE SYMPTOMS OF SLASH-AND-BURN APPROACHES?

There are many possible signs. West (2004) gives the following:

- A process improvement specialist (consultant) cannot tell you who his client is.
- No one can name a single goal for the process improvement effort other than the achievement of a maturity level.
- There is a belief that no processes existed before the CMM initiative.
- Everyone feels like their only job is a maturity level achievement.
  - You are not allowed to use *any* of the old procedures, even ones that worked.
  - Whenever you ask someone "why are you doing that?", they tell you that "the process requires it."
  - People with software delivery responsibilities can recite CMM practices or the identification and titles of their organization's policies and procedures.
  - Estimates for process overhead in development projects exceed 15% of the projects' total effort.
  - The volume of standards and procedures increases, while the quantity and quality of delivered products decrease.
- People use words such as "audit," "inspection," and "compliance."
- People refer to "CMM" or "SEI" requirements.
- People make jokes about the "process police" or "process Gestapo."

7.2.14 What Is the Best Way to Promote Software Quality Improvement without Triggering a Slash-and-Burn Frenzy?

Stelzer and Mellis (1998) provide an excellent set of suggestions of success factors learned from quality improvement initiatives in a large number of organizations that they study. Table 7.1 summarizes these findings.

7.3 SOFTWARE TESTING

7.3.1 What Is the Role of Testing with Respect to Software Quality?

Effective software testing will improve software quality. In fact, even poorly planned and executed testing will improve software quality if it finds defects. Testing is a life cycle activity; testing activities begin from product inception and continue through the delivery of the software and into maintenance. Collecting bug reports and assigning them for repair is also a testing activity. But as a life cycle activity, the most valuable testing activities occur at the beginning of the project. Boehm and Basili (2005) report that finding and fixing a software problem after delivery is often 100 times more expensive than finding and fixing it during the requirements testing phase. And about 40 to 50% of the effort on current software projects is spent on avoidable rework.

7.3.2 Is There a Difference between an Error, a Defect, a Bug, a Fault, and a Failure?

Yes:

- Defect: A flaw in any representation of the software including the requirements, architecture, design, code, or documentation.

TABLE 7.1
Organizational Change in Software Process Improvement

Successful Factor of Organizational Change	Explanation
Change agents and opinion leaders	Change agents initiate and support the improvement projects at the corporate level, opinion leaders at the local level.
Encouraging communication and collaboration	Communication efforts precede and accompany the improvement program (communication) and degree to which staff members from different teams and departments cooperate (collaboration).
Management commitment and support	Management at all organizational levels sponsor the change.
Managing the improvement project	Every process improvement initiative is effectively planned and controlled.
Providing enhanced understanding	Knowledge of current software processes and interrelated business activities is acquired and transferred throughout the organization.
Setting relevant and realistic objectives	Software processes are continually supported, maintained, and improved at a local level.
Stabilizing changed processes	Software processes are continually supported, maintained, and improved at a local level.
Staff involvement	Staff members participate in the improvement activities.
Tailoring improvement initiatives	Improvement efforts are adapted to the specific strengths and weaknesses of different teams and departments.
Unfreezing the organization	The "inner resistance" of an organizational system to change is overcome.

Source: Stelzer & Mellis, 1998.

- Fault: A defect in the software with the potential to cause a failure.
- Error: When an event activates fault in a program.
- Failure: A manifested inability of the program to perform its required functions.
- Bug: A colloquial term for *fault* or *defect*.

7.3.3 Is There a Difference between Verification and Validation?

Verification, or testing, determines whether the products of a given phase of the software development cycle fulfill the requirements established during the previous phase. Verification answers the question "Am I building the product right?"

Validation determines the correctness of the final program or software with respect to the user's needs and requirements. Validation answers the question "Am I building the right product?" Many people refer to "Verification and Validation" (V&V) together because they are closely related activities, and many techniques can be used for both purposes.

7.3.4 WHAT IS THE PURPOSE OF TESTING?

Testing is the execution of a program or partial program with known inputs and outputs that are both predicted and observed in order to find faults or deviations from the requirements. Although testing will flush out errors, this is just one of its purposes. The other is to increase trust in the system. Perhaps once, software testing was thought of as intended to remove all errors. However, testing can only detect the presence of errors, not the absence of them; therefore, it can never be known when all errors have been detected. Instead, testing must increase faith in the system, even though it may still contain undetected faults, by ensuring that the software meets its requirements. This objective places emphasis on solid design techniques and a well-developed requirements document. Moreover, a formal test plan that provides criteria used in deciding whether the system has satisfied the requirements documents must be developed.

7.3.5 WHAT IS A GOOD TEST?

A good test is one that has a high probability of finding an error. A successful test is one that uncovers an error.

7.3.6 WHAT ARE THE BASIC PRINCIPLES OF SOFTWARE TESTING?

The following principles should always be followed (Pressman and Maxim, 2019):

- All tests should be traceable to customer requirements.
- Tests should be planned long before testing begins.
- Remember that the Pareto principle applies to software testing.
- Testing should begin "in the small" and progress toward testing "in the large."
- Exhaustive testing is not practical.
- To be most effective, testing should be conducted by an independent party.

These are the most helpful and practical rules for the tester.

7.3.7 HOW DO I START TESTING ACTIVITIES DURING THE REQUIREMENTS ENGINEERING PROCESS?

Testing is a well-planned activity and should not be conducted unsystematically, nor undertaken at the very end, just as the code is being integrated. The most important activity that the test engineer can conduct during requirements engineering is to ensure that each requirement is testable. A requirement that cannot be tested cannot be guaranteed and, therefore, must be reworked or eliminated. For example, a requirement that says "the system shall be reliable" is untestable. On the other hand, "the MTBF for the system shall be not less than 100 hours of operating time" may be a desirable level of reliability and can be tested; that is, demonstrated to have been satisfied.

There are other testing activities that are then conducted during design and coding.

7.3.8 What Test Activities Occur during Software Design and Code Development?

During the design process, test engineers begin to design the corresponding test cases based on an appropriate methodology. The test engineers and design engineers work together to ensure that features have sufficient testability. Often, the test engineer can point out problems during the design phase, rather than uncover them during testing.

There is a wide range of testing techniques for unit testing, integration testing, and system-level testing. Any one of these test techniques can be either insufficient or computationally unfeasible. Therefore, some combination of testing techniques is almost always employed.

7.3.9 What Different Types of Testing Can Be Performed?

There are three dimensions in the world of testing (see Figure 7.2):

- Dimension 1: Do we consider the internal structure of what we are testing (white box testing) or not (black box testing)?
- Dimension 2: What level are we testing on? Are we testing individual program units, such as procedure, methods, classes in isolation (unit testing); are we testing how few classes are integrated (integration testing); or are we testing the system as a whole (system testing)?
- Dimension 3: What are we testing for? (Usability, performance, security, etc.)

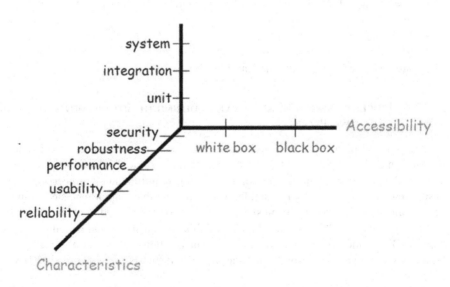

FIGURE 7.2 Dimensions of testing

Any time we conduct any type of testing then we are executing a point in this three-dimensional space. For example:

- We may be testing the whole system to check for performance following black box techniques.
- We may be testing only one class (unit testing) to check for the functionality following a black box technique.

Whatever testing we do; it is always important to relate it to this three-dimensional space. Next, we will look into many techniques to cover the three-dimensional space.

7.3.10 What Are the Different Testing Accessibilities?

We will discuss two categories to generate test cases: black box and white box testing.

7.3.11 What Is Black Box Testing?

In black box testing, only inputs and outputs of the software being tested are considered; how the outputs are generated based on a particular set of inputs is ignored. Such a technique is independent of the implementation of the software.

Here is an example. Suppose we wanted to test a method called "calc_tax" which takes as input a positive real number representing your gross income and outputs the federal and state taxes owed. We can look at the code to determine how it treats different income levels and develop test cases accordingly. On the other hand, if we were given third-party code, some method called "calc_tax2," we could only execute the code (but not see the source). In this case, we might develop the test cases based on equivalence class testing—one class for no income, another for the lowest income tax rate, another for the next, and so on. Or we might do Boundary Value Analysis testing and use test input data that probes around the data boundaries—in this case, formed by the various tax rate levels.

Some widely used black box testing techniques include:

- Exhaustive testing
- Boundary value testing
- Equivalence partitioning
- Worst-case testing
- Random test generation
- Pairwise testing
- Truth Table.

An important aspect of using black box testing techniques is that clearly defined interfaces to the subject being tested are required. We will briefly go over these techniques here, but a more thorough discussion is provided in the book by Jorgensen and DeVries (2021).

7.3.11.1 What Is Exhaustive Testing?

Brute force or exhaustive testing involves presenting each code unit with every possible input combination. Brute force testing can work well in the case of a small number of inputs each with a limited input range; for example, a code unit that evaluates a small number of Boolean inputs. However, a major problem with brute force testing is the combinatorial explosion in the number of test cases. For example, suppose a program takes as input five 32-bit numbers and outputs one 32-bit number. Taking a purely black box testing perspective, to exhaustively test we need to test all possible combinations of five 32-bit inputs. That is we have $2^{32} \cdot 2^{32} \cdot 2^{32} \cdot 2^{32} \cdot 2^{32} = 2^{160}$ test cases. Even if we could randomly generate and run these test cases, say, one every 1 μsec, the entire test set would take more than 4.6×10^{34} years to complete!

7.3.11.2 What Is Boundary Value Testing?

Boundary value or corner case testing solves the problem of combinatorial explosion by testing some very tiny subset of the input combinations identified as meaningful "boundaries" of input.

For example, consider a function that takes as input one variable that lies in the boundary of a certain minimum and maximum values. When generating test cases using BVA technique, the input variable values are used at their min, min+, nom, max−, max. Now, having a function F of x_1 and x_2 (Figure 7.3):

To generate BVA test cases, we have to fix one variable at a nominal value while the changing the value of the other variable to its min, min+, nominal value, max− and max. The test cases will be: $T = \{<x_{1nom}, x_{2min}>, <x_{1nom}, x_{2min}+>, <x_{1nom}, x_{2nom}>, <x_{1nom}, x_{2max}->, <x_{1nom}, x_{2max}+>, <x_{1min}, x_{2nom}>, <X_{1min}+, X_{2nom}>, <x_{1max}-, x_{2nom}>, <x_{1max}, x_{2nom}>\}$ (see Figure 7.3)

Suppose we are to test a program that takes as input three values each represent the length of a side of a triangle where each value is considered valid if it is between 1 and 200. The program is to determine what type of triangle is represented by the sides and output the corresponding determination. Using BVA technique, Table 7.2 presents the generated test cases then:

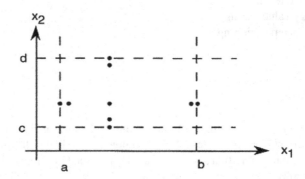

FIGURE 7.3 BVA technique applied on a function with two input parameters

TABLE 7.2
Generated Test Cases from Applying BVA
Technique on the Triangle Program

1 Boundary Value Analysis Test Cases				
Case	a	b	c	Expected Output
1	100	100	1	Isosceles
2	100	100	2	Isosceles
3	100	100	100	Equilateral
4	100	100	199	Isosceles
5	100	100	200	Not a Triangle
6	100	1	100	Isosceles
7	100	2	100	Isosceles
8	100	100	100	Equilateral
9	100	199	100	Isosceles
10	100	200	100	Not a Triangle
11	1	100	100	Isosceles
12	2	100	100	Isosceles
13	100	100	100	Equilateral
14	199	100	100	Isosceles
15	200	100	100	Not a Triangle

A stronger version of this testing could be found if we test values just less than and just greater than each of the boundaries, or we could select more than one nominal value for each input.

7.3.11.3 What Is Equivalence Class Testing?

Equivalence class testing involves partitioning the space of possible test inputs to a code unit or group of code units into a set of representative inputs. We like the analogy of the crash test dummies that auto manufacturers use to ensure the safety of automobiles. Auto manufacturers don't have a crash test dummy representing every possible human being. Instead, they use a few representative dummies—e.g., small, average, and large adult males; small, average, and large adult females; pregnant female; toddler, etc. These categories represent the equivalence classes.

In the same way, we can partition input sets. For example, suppose we are testing a module with input from a sensor that has an expected range of [−1000, 1000]. We can partition this interval in a number of ways. One way might be to consider all the values <−1000 to be in one equivalence class, those in the range [−1000, 1000] in another equivalence class, and those values >1,000 to be a third equivalence class. Then we select a representative input from each of those classes, say −5000, 0, and 5000, and test these cases. We could also combine equivalence class testing with boundary value testing and test the following inputs: −5000, −1000, 0, 1000, and 5000. We can strengthen the testing further by testing around both sides of the boundary values.

7.3.11.4   What Is Worst-Case Testing?

Worst-case, or exception, testing involves those test scenarios that might be considered highly unusual and unlikely. Often these exceptional cases are exactly those for which the code is likely to be poorly designed and, therefore, will fail.

For example, in the baggage inspection system, while it might be highly unlikely that the system is to function at the maximum conveyor speed, this worst-case still needs to be tested.

7.3.11.5   What Is Random Test Case Generation?

Random test case generation, or statistically based testing, can be used for both unit and system-level testing. This kind of testing involves subjecting the code unit to many randomly generated test cases over some period of time. The purpose of this approach is to simulate the execution of the software under realistic conditions.

The randomly generated test cases are based on determining the underlying statistics of the expected inputs. The statistics are usually collected by expert users of similar systems or, if none exist, by informed guessing. The intent of this kind of testing is to simulate typical usage of the system.

The major drawback of such a technique is that the underlying probability distribution functions for the input variables may be unavailable or incorrect. Therefore, randomly generated test cases are likely to miss conditions with low probability of occurrence. These are precisely the kind of situations that may be overlooked in the design of the module. Failing to test these scenarios is an invitation to disaster.

7.3.11.6   What Is Pairwise Testing?

Simply speaking, it is a software testing technique that, for each pair of input parameters to a system (or a method/function/ UI/.), tests all possible discrete combinations of those parameters.

Consider a function with 3 parameters p1, p2, p3 and possible values of 0,1 for each:

- (A) "All values appear once" → two test cases
- (B) "All combinations appear once" → $2^n = 8$ test cases
- (C) Pairwise testing: Only generate all combinations for parameter pairs (P1-P2, P1-P3, P2-P3) and "merge them into test cases". Each possible pair of (Pi-Pj) is included in one of the four All-Pairs test cases (Figure 7.4)

Pairwise testing is a proven industry best practice. The average find-cost per defect for pairwise testing is 0.26 Person/hour for a defect vs. $\approx$ 1 Person/hour for a defect in standard acceptance tests. So, the average is $\approx$ factor 4 better. Pairwise testing generates minimal test case variants that cover all pair combinations'.

The generation of the pair combinations can be a difficult task. Fortunately, there are tools to help; for www.pairwise.org/

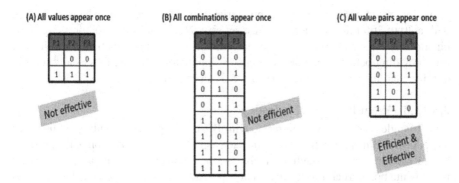

FIGURE 7.4   Comparing "Pairwise Testing" to "All combinations Testing" and "All Values Appear Once"

TABLE 7.3
Animal Traits to Illustrate Combinational Testing

hair=0,	feathers=0	egg-laying=1	milk=0
airborne=0	aquatic=0	predator=0	toothed=0
backbone=1	breathes=1	venomous=0	fins=0
nlegs=4	tail=1	domestic=0	cat size=1

7.3.11.7   What Is Combinatorial Testing?

Combinatorial or t-way testing is a method for more effective testing at lower cost. The key insight underlying its effectiveness resulted from a series of studies by NIST from 1999 to 2004. This research showed that most software faults are caused by one or two parameters, with progressively fewer by three or more. While pairwise testing captures those cases where two parameters may interact to create a fault, combinatorial testing can provide a means when efficient fault detection is needed in cases when three or more parameters may interact to cause the fault.

To briefly illustrate combinatorial testing, consider the animal traits shown in Table 7.3. This table is a tiny subset of the information embedded in the familiar "20 questions" artificial intelligence game implemented via a neural network.

By examination you can see from that table, the 3-way logic shown below is necessary to uniquely identify a reptile.

> not aquatic AND not toothed AND four legs OR egg-laying AND not aquatic AND four legs OR not hairy AND four legs AND cat size OR not milk-producing AND not aquatic AND four legs OR not milk-producing AND four legs AND cat size OR not predator AND not toothed AND four legs

Therefore, we would require cases with these 3-way combinations to assure that this AI worked in the case of identifying a reptile.

Multiple studies have shown combinatorial testing to be equal to exhaustive testing with a 20 to 700 times reduction in test set size. New algorithms compressing combinations into a small number of tests have made this method practical for industrial use, providing better testing at lower cost. Combinatorial testing is typically used in mission- and life-critical systems (Kuhn & Kacker, 2019).

7.3.11.8   What Is Truth Table Testing?

The "triangle problem" we referred to with BVA testing is commonly used in books and papers to illustrate truth table testing, so we will provide a rendition of the problem here. Remember that although truth table testing involves analyzing program logic, it is not white box (structural) testing. That is because the logic that drives the test cases is derived from the specification, not an examination of the code.

Here is the salient part of the specification for the triangle problem. The program is to take three inputs, a, b, and c, which represent the sides of a triangle. The program is to determine what type of triangle is represented by the sides and output the corresponding determination. That is, if all sides are equal, the program outputs "Equilateral"; if exactly one pair of sides is equal, "Isosceles"; and if no pair of sides are equal, "Scalene". Furthermore, if the following conditions are not true:

1. $a < b + c$
2. $b < a + c$
3. $c < a + b$

then the program is to output "Not a Triangle" (you might recall this fact relationship from high school geometry). Let's not worry about any other kind of error or bound checking here—we are only interested in probing the logic of the program.

Now, the requirements create four conditions that are used to determine the output of the program:

c1: the sides represent a triangle, that is, conditions 1, 2, and 3 above are met.
c2: side a = side b
c3: side a = side c
c4: side b = side c

The actions that can be taken by the program are:

a1: output "Not a Triangle"
a2: output "Equilateral"
a3: output "Isosceles"
a4: output "Scalene"

With four logical condition variables (c1 through c4) we should set up a truth table with all possible combinations of those variables—with 16 columns. The top rows represent the conditions and the bottom rows represent those actions that are taken when the corresponding conditions are true. Here is the first attempt at the truth table logic (Table 7.4):

TABLE 7.4
First Attempt at Truth Table for the Triangle Program

Condition	1	2	3	4	5	6	7	8	9	10	11	12	13	14	15	16
c1: a,b,c from a triangle	T	T	T	T	T	T	T	T	F	F	F	F	F	F	F	F
c2: a=b	T	T	T	T	F	F	F	F	T	T	T	T	F	F	F	F
c3: a=c	T	T	F	F	T	T	F	F	T	T	F	F	T	T	F	F
c4: b=c	T	F	T	F	T	F	T	F	T	F	T	F	T	F	T	F
Actions																
Output "Not a Triangle"									T	T	T	T	T	T	T	T
Output "Equilateral"	T															
Output "Isosceles"			T		T	T										
Output "Scalene"							T									

TABLE 7.5
The Resulting Truth Table for the Triangle Program

Condition	1	4	6	7	8	9
c1: a,b,c, from a triangle	T	T	T	T	T	F
c2: a=b	T	T	F	F	F	X
c3: a=c	T	F	T	F	F	X
c4: b=c	T	F	F	T	F	X
Actions						
Output "Not a Triangle"						T
Output "Equilateral"	T					
Output "Isosceles"			T	T	T	
Output "Scalene"					T	

So we have 16 test cases (each column represents a test case with conditions and expected outputs). To avoid clutter in the action section of the table, we only list TRUE values, all other boxes contain FALSE.

Look at this table. First, we note that there are some impossible values in this table. For example, look at column 2. If a=b and b=c are TRUE then how can b=c be FALSE? We have a similar situation in columns 3 and 5. What should we do in these cases? Well, we can't make these situations occur—they are impossible. So, we can remove those so-called test cases. Next, we realize that in columns 9 through 16, the same thing is occurring, that is, as soon as the sides entered fail to represent a triangle (e.g., by putting in a 0 for a side or a negative number) then nothing else matters. Thus, we can collapse those eight columns into a single test case. The resulting truth testing table is shown in Table 7.5.

It has six test cases—the "X" values mean "don't care," that is, either true or false will do. So, we can test this program with six test cases, not 16.

Truth table testing helps us remove redundancy in testing and also identify flaws and redundancy in logic (in the requirements) which will lead to improved program quality.

7.3.12 ARE THERE ANY DISADVANTAGES TO BLACK BOX TESTING?

One disadvantage is that it can bypass unreachable or dead code. In addition, it may not test all of the control paths in the module. In other words, black box testing only tests what is expected to happen, not what wasn't intended. White box or clear box testing techniques can be used to deal with this problem.

7.3.13 WHAT IS WHITE BOX TESTING?

White box testing (sometimes called clear or glass box testing) seeks to test the structure of the underlying code. For this reason, it is also called structural testing.

Whereas black box tests are data-driven, white box tests are logic-driven; that is, they are designed to exercise all paths in the code unit. For example, in the reject mechanism functionality of the baggage inspection system, all error paths would need to be tested including those pathological situations that deal with simultaneous and multiple failures.

White box testing also has the advantage that it can discover those code paths that cannot be executed. This unreachable code is undesirable because it is likely a sign that the logic is incorrect, it wastes code space memory, and it might inadvertently be executed in the case of corruption of the computer's program counter.

The following white box testing strategies will be discussed (this is not an exhaustive list of white box testing techniques, however):

- Control flow testing
- DU path testing
- Code inspection
- Formal program proving.

7.3.13.1 What Is Control Flow Testing?

Control flow testing is a white box testing technique that aims to determine the execution order of statements or instructions of the program through a control structure. The control structure of a program is used to develop a test case for the program. In this technique, a particular part of a large program is selected by the tester to set the testing path. It is mostly used in unit testing (discussed later). Test cases are represented by the control graph of the program.

Here is a simple example. Consider the code fragment below. In order to test the program using the control flow testing technique, the first step is to draw the control flow graph (CFG) that visually reflects the structure of this code. Figure 7.5 shows the corresponding CFG.

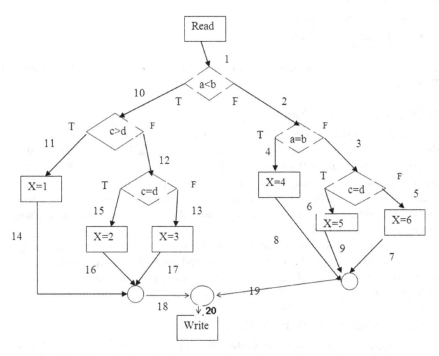

FIGURE 7.5   Control flow graph for code fragment presented above

```
read(a,b,c,d);
if a < b then if c > d then x:= 1 else if c = d then x:= 2
else x:= 3 else if a = b then x:= 4 else if c = d
then x:= 5
else x:= 6;
write(x);
```

A program path is a sequence of statements from entry to exit. There are a total of six different paths from entry to exit in the above code. There is an (input, expected output) pair for each path. Executing a path requires invoking the program unit with the right test input. For example, to execute the path represented by the sequence of <1,10,11,14,18>, then the right test input has to be calculated to force the program into that path when it executes. A possible input for a test case to test the execution of the program into <1,10,11,14,18> path is {a=5, b=7, c=9, d=7}. The expected output is x = 1.

There can be a large number of paths in a program. In the control flow testing technique, paths are chosen by using the concepts of path selection criteria. Examples of these criteria are:

- All Statement Coverage
- All Branch Coverage
- All Path Coverage

- Simple Path Coverage
- Visit Each Loop Once.

For more on control flow testing, interested readers are referred to Jorgensen and DeVries (2021).

7.3.13.2 What Is DU Path Testing?

DU (define-use) path testing is a data-driven white box testing technique that involves the construction of test cases that exercise all possible definition, change, or use of a variable through the software system. Suppose, for example, a variable "acceleration" is defined as a floating-point variable somewhere in the software system but is accessed or changed throughout. Test cases would then be constructed that involve the setting and changing of that variable, and then observing how those changes propagate throughout the system.

DU path testing is actually rather sophisticated because there is a hierarchy of paths involving whether the variable is observed (and, for example, some decision made upon its value) or changed. Interested readers are referred to Jorgensen and DeVries (2021).

7.3.13.3 What Are Code Inspections?

In code inspections, the author of some collection of software presents each line of code to a review group of peer software engineers. Code inspections can detect errors as well as discover ways for improving the implementation. This audit also provides an opportunity to enforce coding standards.

Inspections have been shown to be a very effective form of testing. According to Boehm and Basili (2005), peer code reviews catch 60% of the defects. But when the reviews are directed (meaning, the reviewers are asked to focus on specific issues), then 35% more defects are caught than in non-directed reviews.

7.3.13.4 What Is Formal Program Proving?

Formal program proving is a kind of white box testing using mathematical techniques in which the code is treated as a theorem and some form of calculus is used to prove that the program is correct. This form of verification requires a high level of training and is useful, generally, for only limited purposes because of the intensity of activity required.

7.3.14 WHAT ARE THE DIFFERENT TESTING LEVELS?

There are many levels of testing; for example, testing can be conducted at the levels of:

- Unit testing: Individual program units, such as procedure, methods in isolation
- Integration testing: Modules are assembled to construct a larger subsystem and tested
- System testing: Includes a wide spectrum of testing such as functionality, and load

- Acceptance testing: Customer's expectations from the system. There are different types of acceptance testing; for example:
  - User acceptance testing (UAT): The system satisfies the contractual acceptance criteria
  - Business acceptance testing (BAT): System will eventually pass the user acceptance test.

7.3.15 What Is Unit-Level Testing?

Several methods can be used to test individual modules or units. These techniques can be used by the unit author (sometimes called desk checking) and by the independent test team to exercise each unit in the system. These techniques can also be applied to subsystems (collections of modules related to the same function). The techniques to be discussed include black box and white box testing.

7.3.16 What Is System Integration Testing?

Integration testing involves the testing of groups of components integrated to create a system or subsystem. The tests are derived from the system specification. The principal challenge in integration testing is locating the source of the error when a test fails. Incremental integration testing reduces this problem.

Once individual modules have been tested, then subsystems or the entire system need to be tested. In larger systems, the process can be broken down into a series of subsystem tests and then a test of the overall system.

If an error occurs during system-level testing, the error must be repaired. Ideally, every test case involving the changed module must be rerun, and all previous system-level tests must be passed in succession. The collection of system test cases is often called a system test suite.

7.3.17 What Is Incremental Integration Testing?

This is a strategy that partitions the system in some way to reduce the code tested. The incremental testing strategy includes:

- Top-down testing
- Bottom-up testing
- Sandwich integration
- Cleanroom testing
- Interface testing

In practice, most integration involves a combination of these strategies.

7.3.18 What Is the Difference between Top-Down and Bottom-Up Testing?

Consider Figure 7.6, which represents partial top-down testing of some fictitious code. The nodes represent code units and the directed arrows represent a calling or

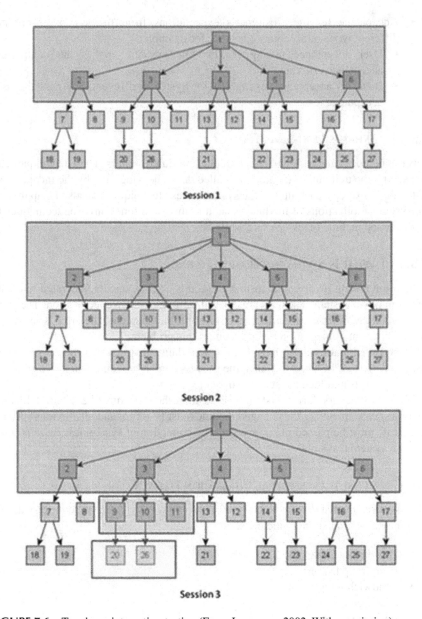

FIGURE 7.6   Top-down integration testing (From Jorgensen, 2002. With permission)

invocation (if they are methods) sequence between those code units. The shaded areas represent the collection of code units to be tested in the appropriate test cases, organized as test sessions.

The testers work their way down from the "top" of the system, which is the main program if it is written in a procedural manner. If it is an object-oriented program, then it is much trickier to identify the sequence of method invocations.

In session 1 stubs need to be written to simulate interfaces at code units 7 through 17. In session 2 stubs also need to be written at nodes 20 and 26. In session 3 no new stubs need to be written. The stubs would have the appropriate parameter interface but do nothing, except perhaps indicating somehow that they were successfully called or invoked.

If we were to test this software in a bottom-up fashion, the testing sessions are shown in Figure 7.7.

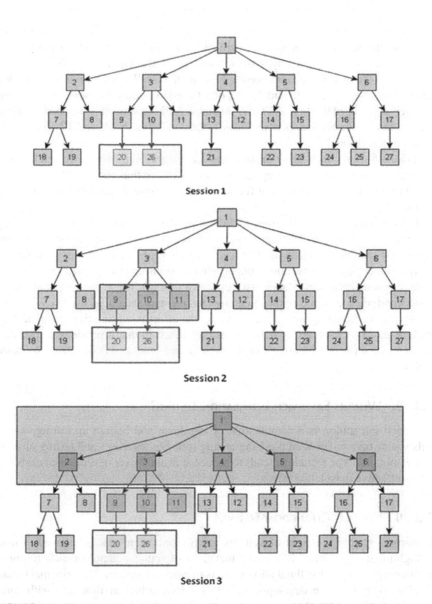

FIGURE 7.7   Bottom-up integration testing (From Jorgensen, 2002. With permission)

TABLE 7.6
Number of Stubs and Drivers from Top-Down and Bottom-Up Sessions Presented in Figures 7.6 and 7.7

Integration Testing Approach	Stubs Needed	Drivers Needed
Top-down	13	0
Bottom-up	8	5

While the sessions look similar, here the testing sequence starts at the bottom level subtree and works its way to the top subtree. In this case, drivers (which simulate calls to the modules below) need to be written as well as stubs. For session 1 we need drivers for modules 20 and 26. We don't need any stubs. For session 2 we need a driver at modules 9, 10, and 11 and no stubs. And for session 3 we need no more drivers, but we do need stubs at modules 7, 8, 13, 12, 14, 15, 16, and 17. This yields a total of 8 stubs and 5 drivers.

Table 7.6 shows the number of stubs and drivers needed for top-down and bottom-up sessions if we were to complete bottom-up integration testing. The stubs and drivers needed would be different for sandwich, pairwise, or neighborhood integration testing.

It takes time to write the stubs and drivers, which are essentially thrown away when testing is finished. This throwaway code is often not accounted for in software project estimation and planning. Further, there are effort trade-offs between throwaway code, testing low-level functionality first (bottom-up testing), and testing high-level functionality and interfaces first (top-down testing). These trade-offs need to be understood prior to selecting the integration testing approach to be used.

No-one has come up with a way to assess the effort in writing throwaway testing code. A paper by Morozoff (2010) comes the closest in that it provides a good discussion of measuring the effects of throwaway and reworked code during software maintenance.

7.3.19 What Is Sandwich Integration Testing?

Sandwich integration is a combination of top-down and bottom-up testing, which falls somewhere in between big-bang testing (one big test case) and testing of individual modules. The technique tends to reduce stub and driver development costs but is less effective in isolating errors.

7.3.20 What Is Cleanroom Testing?

Cleanroom testing is a kind of systems integration testing that is believed to lead to high-integrity systems. It uses the metaphor of semiconductor manufacturing—creating a process that will not allow defects to enter the system. The principal tenant of cleanroom software development is that, given sufficient time and with care, error-free software can be written. Cleanroom software development relies heavily on group walkthroughs, code inspections, code reading by stepwise abstraction, and formal program validation. It requires that software specifications exist that are sufficient to completely describe the system.

In the strictest forms of this approach, the development team is not allowed to test code as it is being developed. Rather, syntax checkers, code walkthroughs, group inspections, and formal verifications are used to ensure software integrity. The program is developed by slowly integrating features into the code, starting with some baseline of functionality. At each milestone, an independent test team checks the code against a set of randomly generated test cases based on a set of statistics describing the frequency of use for each feature specified in the requirements.

This group tests the code incrementally at predetermined milestones, and either accepts or returns it to the development team for correction. Once a functional milestone has been reached, the development team adds to the "clean" code, using the same techniques as before. Thus, like an onion skin, new layers of functionality are added to the software system unit it has completely satisfied the requirements.

Numerous successful projects have been developed using strict or modified cleanroom testing in both academic and industrial environments. In any case, many of the tenants of cleanroom testing can be incorporated without completely embracing the methodology (Laplante, 2004).

7.3.21 WHAT IS INTERFACE TESTING?

This testing takes place when modules or subsystems are integrated to create larger systems. Therefore, it can take place during integration testing. The objective is to detect faults due to interface errors or invalid assumptions about interfaces. This kind of testing is particularly important for object-oriented development as objects are defined by their interfaces.

7.3.22 WHAT KINDS OF INTERFACES CAN BE TESTED?

Parameter interfaces can be tested to ensure that the correct data is passed from one procedure to another. Shared memory interfaces can be tested by reading and writing global variables from all code units that can access those variables. Procedural interfaces can be tested to verify that each subsystem encapsulates a set of procedures to be called by other subsystems. Finally, message-passing interfaces can be tested to verify that subsystems correctly request services from other subsystems.

7.3.23 WHAT KINDS OF ERRORS CAN OCCUR AT THE INTERFACES?

One error is interface misuse, where a calling component calls another component and makes an error in its use of its interface; for example, when parameters are in the wrong order. Another type of error is interface misunderstanding, which occurs when a calling component embeds incorrect assumptions about the behavior of the called component. Finally, timing errors can occur at the interfaces in that the called and the calling component operate at different speeds and out-of-date information is accessed.

7.3.24 What Are Some Guidelines for Testing Interfaces?

First, design tests so that parameters to a called procedure are at the extreme ends of their ranges. Then, always test pointer parameters with null pointers. Use stress testing (to be discussed shortly) in message-passing systems and in shared memory systems, and vary the order in which components are activated. Finally, design tests that cause the component to fail.

7.3.25 What Are System and Acceptance Testing?

Near the end of the software life cycle we are interested in demonstrating that the software meets its requirements—this is the verification process and it answers the question "did we build the software right?" Thus, our test case design at this point is focused on the functional and non-functional requirements of the requirements specification.

Sometimes, however, we don't have a requirements specification for the software (in the case of open-source or externally furnished software) or we don't believe that the requirements specification is correct (in the case of some legacy software). In these cases, we will need to build a representation of the requirements specification called a behavioral specification.

There are many kinds of systems tests that probe all kinds of functional requirements and non-functional requirements. These elements of the characteristics dimension from Figure 7.2 are discussed next.

7.3.26 What Are the Different Characteristics to Test For?

Naik and Tripathy (2008) presented a taxonomy of systems testing. The different types of testing discussed in the taxonomy include:

- *Basic tests* provide evidence that the system can be installed, configured, and be brought to an operational state.
- *Functionality tests* provide comprehensive testing over the full range of the requirements, within the capabilities of the system.
- *Robustness tests* determine how well the system recovers from various input errors and other failure situations.
- *Interoperability tests* determine whether the system can inter-operate with other third-party products.
- *Performance tests* measure the performance characteristics of the system, e.g., throughput and response time, under various conditions.
- *Scalability tests* determine the scaling limits of the system, in terms of user scaling, geographic scaling, and resource scaling.
- *Stress tests* put a system under stress in order to determine the limitations of a system and, when it fails, to determine the manner in which the failure occurs.
- *Load and stability* tests provide evidence that the system remains stable for a long period of time under full load.

- *Reliability tests* measure the ability of the system to keep operating for a long time without developing failures.
- *Regression tests* determine that the system remains stable as it cycles through the integration of other subsystems and through maintenance tasks.
- *Documentation tests* ensure that the system's user guides are accurate and usable.

7.3.27 What Is Object-Oriented Testing?

Object-oriented testing is an approach that focuses on objects attributes and their interactions as the bases of testing, rather than on control flow and function execution. Object-oriented testing is most appropriate when systems are designed using an object-oriented approach and/or when the programming language used is object-oriented.

In object-oriented testing the components to be tested are object classes that are instantiated as objects. Because the interactions among objects have a larger grain than do individual functions, systems integration testing approaches have to be used. But the problem is further complicated because there is no obvious "top" to the system for top-down integration and testing.

The three testing levels are:

- Object classes
- Clusters of cooperating objects
- The complete object-oriented system

7.3.28 How Are Object Classes Tested?

Inheritance makes it more difficult to design object class tests, as the information to be tested is not localized. Object class testing can be achieved by

- Testing all methods associated with an object
- Setting and interrogating all object attributes
- Exercising the object in all possible states

Note that methods can be tested using any of the black or white box testing techniques discussed for unit testing.

7.3.29 How Can Clusters of Cooperating Objects Be Tested?

Various techniques are used to identify clusters of objects using knowledge of the operation of objects and the system features that are implemented by these clusters. Cluster identification approaches include:

- Use case or scenario testing
- Thread testing

- Object interaction testing
- Uses-based testing

Use case or scenario testing is based on a user interaction with the system. This kind of cluster testing has the advantage that it tests system features as experienced by users.

Thread testing focuses on each thread. A thread is all of the classes needed to respond to a single external input. Each class is unit tested, and then the thread set is exercised.

Object interaction testing tests sequences of object interactions that stop when an object operation does not call on services from another object.

Uses-based testing begins by testing classes that use few or no server classes. It continues by testing classes that use the first group of classes, followed by classes that use the second group, and so on.

7.3.30   What Is Scenario Testing?

This type of system testing for object-oriented systems involves identifying scenarios from use cases and supplementing them with interaction diagrams that show the objects involved in the scenario.

7.3.31   What Is Software Fault Injection?

Fault injection is a form of dynamic software testing that acts like "crash-testing" the software by demonstrating the consequences of incorrect code or data. Anyone who has ever tried to type a letter when the input called for a number is familiar with fault injection.

The main benefit of fault injection testing is that it can demonstrate that the software is unlikely to do what it shouldn't. Fault injection can also help to reveal new output states that have never before been observed or contemplated.

Fault injection can also be used as a test stoppage criterion; for example, test until fault injection no longer causes failure. Finally, it can be used as a safety case proposition—"Hey, we tested this system and injected all kinds of crazy faults and it didn't break" (Voas, 1998).

7.3.32   When Should You Stop Testing?

There are several criteria that can be used to determine when testing should cease. These include:

- When you run out of time
- When continued testing causes no new failures
- When continued testing reveals no new faults
- When you can't think of any new test cases
- When you reach a point of "diminishing returns"
- When mandated coverage has been attained
- When all faults have been removed.

(Jorgensen & DeVries, 2021)

But the best way to know when testing is complete is when the test coverage metric requirements have been satisfied.

7.3.33 WHAT ARE TEST COVERAGE METRICS?

Test coverage metrics are used to characterize the test plan and identify insufficient or excessive testing. Two simple metrics are requirements per test and tests per requirement. Obviously, every requirement should have one or more test associated with it. But every test, to be efficient, should test more than one requirement. However, one would not want an average of one test per requirement because this is essentially "big-bang" testing, and is ineffective in uncovering errors. Similarly, if a test covers too many requirements and it fails, it might be hard to localize the error.

Research is always ongoing to determine appropriate values for these statistics. But in any case, you can look for inconsistencies to determine if some requirements are not being tested thoroughly enough, if some requirements need more testing, and if some tests are too complex. Remember, if a test covers too many requirements and fails, it could be difficult to localize the error. There is always a trade-off between the time and cost of testing vs. the comprehensiveness of testing.

7.3.34 HOW DO I WRITE A TEST PLAN?

The test plan should follow the requirements document item by item, providing criteria that are used to judge whether the required item has been met. A set of test cases is then written which is used to measure the criteria set out in the test plan. Writing such test cases can be extremely difficult when a user interface is part of the requirements.

The test plan includes criteria for testing the software on a module-by-module or unit level, and on a system or subsystem level; both should be incorporated in a good testing scheme. The system-level testing provides criteria for the hardware/software integration process. IEEE Standard 829–2008 (IEEE Standard for Software Test Documentation) can be helpful for those unfamiliar with software documents.

7.3.35 ARE THERE AUTOMATED TOOLS FOR TESTING THAT CAN MAKE THE JOB EASIER?

Software testing can be an expensive proposition to conduct, but well worth the cost if done correctly. Testing workbenches provide a range of tools to reduce the time required and total testing costs. Most testing workbenches are open systems because testing needs are organization specific.

7.3.36 WHAT ARE SOME TESTING TOOLS THAT I CAN USE?

Mastery of testing tools is a sought-after competency by employers for software testing positions. According to a recent study by Kassab et al. (2021) that analyzed 1,000 job ads related to software testing positions, 56% of them asked for testing tool skills as a requirement or as a preference. Most of the job posts advertised for software testers who had mastered at least one testing tool, and less than half ask for at least specific two tools. Only 14% of the overall ads ask for competency in five or more tools. The average number of tools per ad among those asking for tools is 3.

The testing tool the employers are most interested in is Selenium (Table 7.7). This is not surprising considering the high demand for automation and that Selenium doesn't require sophisticated technical skills. Selenium provides a playback to authored functional tests for web applications without the need to program using a test scripting language. Cucumber is the second most desired in the automating testing tools category. This tool executes automated acceptance tests written in a behavior-driven development. This allows specifying expected software behaviors in a logical language that is more understandable to customers. QTP is another functional testing automation tool in demand by employers.

The tools in the test management category were slightly less in demand than the test execution tools. Jira is the most frequently demanded tool for this category and it incorporates a variety of add-on features for bug-tracking, management and integration, and time tracking for projects. The results from the study by Kassab et al. (2021) shows that performance was the most tested quality attribute. JMeter and LoadRunner are in demand in this category and are both performance measurement tools that were previously within the responsibility of developers but are currently

TABLE 7.7
Demands for Tools Used in Testing

Tool	Purpose	Reported Usage
Selenium	Automates test execution	43.2%
Jira	Test management, bug tracking	38%
Cucumber	Automates test execution	9.5%
SoapUI	Automates test execution, API testing	8.5%
Jmeter	Performance Testing	7.8%
Team Foundation Server	Test management	7.2%
HP QC\ALM	Test management	6.3%
LoadRunner	Performance Testing	6.2%
Confluence	Test management	5.8%
TestNG	Automates test execution, unit testing	5.8%
HP QTP	Automates test execution	5.5%
TestRail	Test management	5.5%
JUnit	Unit testing	4.4%
Katalon Studio	Automates test execution	2.65%
Bugzilla	Bug tracking	1.94%
TestComplete	Automates test execution	1.94%
Microsoft Test Manager	Test management	1.76%
Ranorex	Automates test execution	1.41%
Tosca	Automates test execution	1.23%

Source: Kassab et al., 2021.

reaching the territory of testers' competencies as well. Although a significant number of companies employ bug-tracking tools, only a few job posts explicitly require competency in these tools as shown in Table 7.7.

 Share your opinion: From your professional experience, what testing tools are you mostly using to test a software?
https://pennstate.qualtrics.com/jfe/form/SV_6hvpjqDNq10YJqC

7.3.37 WHAT IS THE CURRENT STATE OF PRACTICE IN SOFTWARE TESTING IN THE INDUSTRY?

In Kassab et al. (2017), the authors presented the results from a web-based survey that examined how software professionals used testing. The results offer opportunities for further interpretation and comparison to software testers, project managers, and researchers. The data includes characteristics of practitioners, organizations, projects, and practices. The interested readers may refer to the study for a comprehensive overview of the testing state of practice.

7.4 METRICS

7.4.1 WHAT ARE SOME MOTIVATIONS FOR MEASUREMENT?

The key to controlling anything is measurement. Software is no different in this regard, but the question arises "what aspects of software can be measured?" Chapter 2 introduced several important software properties and alluded to their measurement. It is now appropriate to examine the measurement of these properties and show how this data can be used to monitor and manage the development of software.

Metrics can be used in software engineering in several ways. First, certain metrics can be used during software requirements development to assist in cost estimation. Another useful application for metrics is benchmarking. For example, if a company has a set of successful systems, then computing metrics for those systems yields a set of desirable and measurable characteristics with which to seek or compare in future systems.

Most metrics can also be used for testing in the sense of measuring the desirable properties of the software and setting limits on the bounds of those criteria. Or they can be used during the testing phase and for debugging purposes to help focus on likely sources of errors.

Of course, metrics can be used to track project progress. In fact, some companies reward employees based on the amount of software developed per day as measured by some of the metrics to be discussed (e.g., delivered source instructions, function points, or lines of code).

Finally, as Kelvin's quote at the start of chapter 2 suggests, a quality culture depends upon measurement. Such an approach is similar to other kinds of engineering.

7.4.2 WHAT KINDS OF METRICS EXIST FOR TRACKING SYSTEM TESTS?

We categorize execution metrics into two classes:

- Metrics for monitoring test execution
- Metrics for monitoring defects

7.4.3 WHAT ARE THE METRICS FOR MONITORING TEST EXECUTION?

Sample of these metrics include:

- Test Case Escapes (TCE): This metric measures the number of defects that were found outside the scope of the existing test cases. A significant increase in the number of test case escapes implies deficiencies in the test design.
- Planned versus Actual Execution (PAE) Rate: This compares the actual number of test cases executed every week with the planned number of test cases.
- Execution Status of Test (EST) Cases: Periodically monitors the number of test cases lying in different states: Failed, Passed, Blocked, Invalid and Untested). This metric is useful to further subdivide those numbers by test categories.

7.4.4 WHAT ARE THE METRICS FOR MONITORING DEFECTS?

Sample of these metrics include:

- Irreproducible defects (IRD) count
- Defects arrival rate (DAR) count
- Defects rejected rate (DRR) count
- Outstanding defects (OD) count
- Crash defects count
- Arrival and resolution of defects (ARD) count
- Defects classified as function as designed (FAD) count

Another useful metric is to measure the Defect Removal Efficiency (DRE). DRE can be calculated using the following formula:

$$\frac{Number\ of\ Defects\ Found\ in\ Testing}{Number\ of\ Defects\ Found\ in\ Testing + Number\ of\ Defects\ Not\ Found} \quad (7.1)$$

Other useful emerging metrics include the spoilage metric, to be discussed next.

7.4.5 WHAT IS THE SPOILAGE METRIC?

Defects are injected and removed at different phases of a software development cycle. The cost of each defect injected in phase X and removed in phase Y increases with the increase in the distance between X and Y. An effective testing method would find defects earlier than a less effective testing method would. A useful measure of test effectiveness is defect age, called PhAge, which measures the number of phases between defect injection and defect removal. The spoilage metric can be calculated by the formula:

$$Spoilage = \frac{\sum(Number\ of\ Defects*Discovered\ Phage)}{Total\ Number\ of\ Defects} \quad (7.2)$$

A spoilage value close to 1 is an indication of a more effective defect discovery process. As an absolute value, the spoilage metric has little meaning. This metric is useful in measuring the long-term trend of test effectiveness in an organization.

7.4.6 IN ADDITION TO TEST EXECUTION METRICS, WHAT OTHER KINDS OF THINGS CAN WE MEASURE IN SOFTWARE?

We can measure many things. Typical candidates include:

- Lines of code
- Code paths
- Change rates
- Elapsed project time
- Budget expended

7.4.7 IS THE "LINES OF CODE" METRIC USEFUL?

The easiest characteristic of software that can be measured is the number of lines of finished source code. Measured as thousands of lines of code (KLOC), the "clock" metric is also referred to as delivered source instructions (DSI) or non-commented source code statements (NCSS). That is, we count executable program instructions (excluding comment statements, header files, formatting statements, macros, and anything that does not show up as executable code after compilation or cause allocation of memory).

Another related metric is source lines of code (SLOC), the major difference being that a single source line of code may span several lines. For example, an if-then-else statement would be a single SLOC, but many DSI.

While the KLOC metric essentially measures the weight of a printout of the source code, thinking in these terms makes it likely that the usefulness of KLOC will be unjustifiably dismissed as supercilious. But isn't it likely that 1,000 lines of code is going to have more defects than 100 lines of code? Wouldn't it take longer to develop the latter than the former? Of course, the answer is dependent upon the complexity of the code.

7.4.8 WHAT ARE THE DISADVANTAGES OF THE LOC METRIC?

One of the main disadvantages of using lines of source code as a metric is that it can only be measured after the code has been written. While lines of code can be estimated, this approach is far less accurate than measuring the code after it has been written.

Another criticism of the KLOC metric is that it does not take into account the complexity of the software involved. For example, 1,000 lines of print statements probably do have fewer errors than 100 lines of a complex imaging algorithm. Nevertheless, KLOC is a useful metric, and in many cases is better than measuring nothing. Many other metrics are fundamentally based on lines of code.

7.4.9 WHAT ARE THE DELTA LINES OF CODE METRIC?

Delta KLOC measures how many lines of code change over some period of time. Such a measure is useful, perhaps, in the sense that as a project nears the end of code development, Delta KLOC would be expected to be small. Other, more substantial, metrics are also derived from KLOC.

7.4.10 WHAT IS MCCABE'S METRIC?

To attempt to measure software complexity, McCabe (1976) introduced a metric, cyclomatic complexity, to measure program flow-of-control. This metric is one of the most important and long-lasting concepts in software engineering. While the concept of creating metrics based on control flow fits well with procedural programming it does not necessarily comport with object-oriented programming. There are adaptations, however, that are regularly used for object-oriented systems, which we will discuss later. In any case, McCabe's metric has two primary uses:

1. To indicate escalating complexity in a module as it is coded and therefore assisting the coders in determining the "size" of their modules
2. To determine the upper bound on the number of tests that must be designed and executed

The use of the McCabe metric in calculating the minimum number of test cases needed to traverse all linearly independent code paths was already discussed.

7.4.11 HOW DOES MCCABE'S METRIC MEASURE SOFTWARE COMPLEXITY?

The cyclomatic complexity is based on determining the number of linearly independent paths in a program module, suggesting that the complexity increases with this number and reliability decreases.

To compute the metric, the following procedure is followed. Consider the flow graph of a program. Let e be the number of edges and n be the number of nodes. Form the cyclomatic complexity, C, as follows:

$$C = e - n + 2 \qquad (7.3)$$

This is the most generally accepted form.[1]

7.4.12 CAN YOU HELP ME VISUALIZE THE CYCLOMATIC COMPLEXITY?

To get a sense of the relationship between program flow and cyclomatic complexity, refer to Figure 7.8. Here, for example, a sequence of instructions has two nodes, one

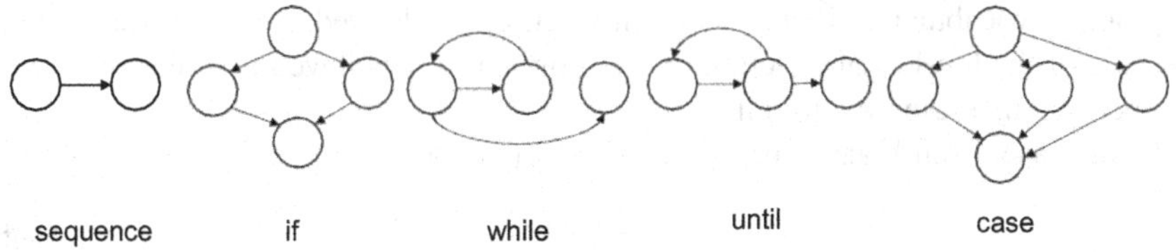

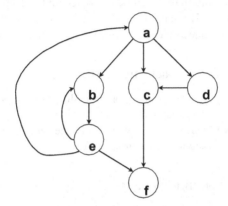

FIGURE 7.8 Correspondence of language statements and flow graph

sequence   if   while   until   case

FIGURE 7.9 Flow graph for noise reduction code for the baggage inspection system

edge, and one region and, hence, would contribute a complexity of $C = 1 - 2 + 2 = 1$. This is intuitively pleasing as nothing could be less complex than a simple sequence.

On the other hand, the case statement, which has six edges, five nodes, and two regions, would contribute $C = 6 - 5 + 2 = 3$ to the overall complexity.

As a more substantial example, consider a segment of code extracted from the noise reduction portion of the baggage inspection system. The procedure calls between modules **a**, **b**, **c**, **d**, **e**, and **f** are depicted in Figure 7.9.

Here, then, $e = 9$ and $n = 6$ yield a cyclomatic complexity of $C = 9 - 6 + 2 = 5$.

7.4.13 CAN THE COMPUTATION OF MCCABE'S METRIC BE AUTOMATED?

Computation of McCabe's metric can be done easily during compilation by analyzing the internal tree structure generated during the parsing phase. However, commercial tools are available to perform this analysis.

7.4.14 WHAT ARE HALSTEAD'S METRICS?

Halstead's metrics measure information content, or how intensively the programming language is used. Halstead's metrics are based on the number of distinct, syntactic elements and begin–end pairs (or their equivalent, such as open and closed curly braces in Java or C). From these, a statistic for program length is determined. We will omit the equations because they are rarely computed by hand. From these statistics, a "program vocabulary," V, and program level, L, are derived. L is supposed to be a measure of the level of abstraction of the program. It is believed that increasing this number will increase system reliability.

In any case, from V and L the effort, E, is defined as

$$E = V/L \qquad (7.4)$$

Decreasing the effort level is believed to increase reliability as well as ease of implementation.

In principle, the program length can be estimated and, therefore, is useful in cost and schedule estimation. The length is also a measure of the "complexity" of the program in terms of language usage and, therefore, can be used to estimate defect rates.

7.4.15 ARE HALSTEAD'S METRICS STILL USED?

Halstead's metrics, though dating back more than 40 years, are still widely used and tools are available to automate calculations. Interestingly, because they are language-independent, they have been used in a variety of other applications beyond testing, notable, for plagiarism detection in program code.

7.4.16 WHAT ARE FUNCTION POINTS?

Function points (FPs) were introduced in the late 1970s as an alternative to metrics based on simple source line count. The basis of FPs is that, as more powerful programming languages are developed, the number of source lines necessary to perform a given function decreases. Paradoxically, however, the cost per LOC measure indicated a reduction in productivity, as the fixed costs of software production were largely unchanged.

The solution to this effort estimation paradox is to measure the functionality of software via the number of interfaces between modules and subsystems in programs or systems. A big advantage of the FP metric is that it can be calculated before any coding occurs.

7.4.17 WHAT ARE THE PRIMARY DRIVERS FOR FPS?

The following five software characteristics for each module, subsystem, or system represent its FPs:

- Number of inputs to the application (I)
- Number of outputs (O)
- Number of user inquiries (Q)
- Number of files used (F)
- Number of external interfaces (X).

In addition, the FP calculation takes into account weighting factors for each aspect that reflects their relative difficulty in implementation. These coefficients vary depending on the type of application system. Then complexity factor adjustments can be applied for different types of application domains. The full set of coefficients and corresponding questions can be found by consulting an appropriate text on software metrics.

7.4.18   How Do I Interpret the FP Value?

Intuitively, the higher the FP, the more difficult the system is to implement. For the purposes of comparison, and as a management tool, FPs have been mapped to the relative lines of source code in particular programming languages. These are shown in Table 7.8.

For example, it seems intuitively pleasing that it would take many more lines of assembly language code to express functionality than it would in a high-level language like C. In the case of the baggage inspection system, with FP = 241, it might be expected that about 31,000 LOC would be needed to implement the functionality. In turn, it should take many less LOC to express that same functionality in a more abstract language such as C++. The same observations that apply to software production might also apply to maintenance as well as to the potential reliability of software.

7.4.19   How Widely Is the FP Metric Used?

The FP metric is widely used in business applications, but not nearly as much in embedded systems. However, there is increasing interest in the use of FPs in real-time embedded systems, especially in large-scale real-time databases, multimedia, and Internet support. These systems are data-driven and often behave like the large-scale transaction-based systems for which FPs were developed.

The International Function Point Users Group maintains a Web database of weighting factors and FP values for a variety of application domains. These can be used for comparison.

TABLE 7.8
Programming Language and Lines of Code per FP

Language	LOC per FP
Assembly	320
C	128
Fortran	106
C++	64
Visual Basic	32
Smalltalk	22
SQL	12

Source: Adapted from Jones, 2007.

7.4.20 What Are Feature Points?

Feature points are an extension of FPs developed by Software Productivity Research, Inc. in 1986. Feature points address the fact that the classical FP metric was developed for management information systems and, therefore, is not particularly applicable to many other systems, such as real-time, embedded, communications, and process control software. The motivation is that these systems exhibit high levels of algorithmic complexity, but sparse inputs and outputs.

The feature point metric is computed in a manner similar to the FP except that a new factor for the number of algorithms, A, is added.

7.4.21 Are There Special Metrics for Object-Oriented Software?

While any of the previously discussed metrics can be used in object-oriented code, other metrics are better suited for this setting. Object-oriented metrics are computed on three levels:

- Methods
- Classes
- Packages

7.4.22 What Are Some Method-Level Metrics?

Often the lines of code, McCabe cyclomatic complexity, or even Halstead's metrics are used.

7.4.23 What Are Commonly Used Class-Level Metrics?

The most widely used set of metrics for the object-oriented software, the Chidamber and Kemerer (C&K) metrics, contains metrics that are at the class level. These are primarily applied to the concepts of classes, coupling, and inheritance.

Some of the C&K metrics are:

- Weighted Methods per Class (WMC): The sum of the complexities of the methods (method complexity is measured by cyclomatic complexity, CC).
- Response for a Class (RFC): The number of methods that can be invoked in response to a message to an object of the class or by some method in the class. Includes all methods accessible within the class hierarchy.
- Lack of Cohesion in Methods (LCOM): The dissimilarity of methods in a class by instance variable or attributes.
- Coupling between Object Classes (CBO): The number of other classes to which a class is coupled. Measured by counting the number of distinct non-inheritance-related class hierarchies on which a class depends.
- Depth of Inheritance Tree (DIT): The maximum number of steps from the class node to the root of the tree. Measured by the number of ancestor classes.
- Number of Children (NOC): The number of immediate subclasses subordinate to a class in the hierarchy.

7.4.24 WHAT ARE SOME PACKAGE-LEVEL METRICS?

Martin's package cohesion metrics are based on the principles of good OOD, as discussed in Chapter 4.

- Afferent Coupling (A): The number of other packages that depend upon classes within this package. This metric is an indicator of the package's responsibility.
- Efferent Coupling: The number of other packages that the classes in this package depend upon. This metric is an indicator of the package's independence.
- Abstractness: The ratio of the number of abstract classes (and interfaces) in this package to the total number of classes in the analyzed package.
- Instability (I): The ratio of efferent coupling to total coupling.
- Distance from the Main Sequence: The perpendicular distance of a package from the idealized line, $A + I = 1$.
- Law of Demeter: This can be informally stated as "only talk to your friends," that is, limit the intellectual distance from any object and anything it uses. This metric is measured by the number of methods called subsequently by the external objects participating in the method chain.

7.4.25 ARE THERE OTHER KINDS OF OBJECT-ORIENTED METRICS?

Yes. There seem to be more metrics for object-oriented code than for procedural code. Other object-oriented metrics track things like component dependencies or rely on various features (e.g., depth, cycles, or fullness) of graphs depicting packages or classes relationships.

7.4.26 WHAT ARE OBJECT POINTS?

This is a function point-like metric and is not specifically intended for use with object-oriented code, despite its name. Like FPs, it is a weighted estimate of the following visible program features:

- Number of separate screens
- Number of reports
- Number of third-generation language modules needed to support fourth-generation language code.

Object points are an alternative to FPs when fourth-generation languages are used.

7.4.27 WHAT ARE USE CASE POINTS?

Use case points (UCPs) allow the estimation of an application's size and effort from its use cases. UCPs are based on the number of actors, scenarios, and various technical and environmental factors in the use case diagram. The UCP equation is based on four variables:

- Technical complexity factor (TCF)
- Environment complexity factor (ECF)
- Unadjusted use case points (UUCP)
- Productivity factor (PF).

which yield the equation:

$$UCP = TCP * ECF * UUCP * PF \qquad (7.5)$$

UCPs are a relatively new estimation technique.

7.4.28   THIS IS ALL SO CONFUSING. WHICH METRICS SHOULD I USE?

The goal question metric (GQM) paradigm is a helpful framework for selecting the appropriate metrics.

7.4.28.1   What Is the GQM Technique?

GQM is an analysis technique that helps in the selection of an appropriate metric. To use the technique, you follow three simple rules. First, state the goals of the measurement; that is, what is the organization trying to achieve? Next, derive from each goal the questions that must be answered to determine if the goals are being met. Finally, decide what must be measured in order to be able to answer the questions.

7.4.28.2   Can You Give a Simple Example?

Suppose one of your organization's goals is to evaluate the effectiveness of the coding standard. Some of the associated questions you might ask to assess if this goal has been achieved are:

- Who is using the standard?
- What is coder productivity?
- What is the code quality?

For these questions, the corresponding metrics might be:

- What proportion of coders is using the standard?
- How has the number of LOC or FPs generated per day per coder changed?
- How have appropriate measures of quality for the code changed? (Appropriate measures might be errors found per LOC or cyclomatic complexity.)

Now that this framework has been established, appropriate steps can be taken to collect, analyze, and disseminate data for the study.

7.4.29   WHAT ARE SOME OBJECTIONS TO USING METRICS?

Some widely used objections are that metrics can be misused or that they are a costly and unnecessary distraction. For example, metrics related to the number of

LOC imply that the more powerful the language, the less productive the programmer. Hence, obsessing with code production based on LOC is a meaningless endeavor.

Another objection is that the measuring of the correlation effects of a metric without clearly understanding the causality is unscientific and dangerous. For example, while there are numerous studies suggesting that lowering the cyclomatic complexity leads to more reliable software, there just isn't any real way to know why. Obviously, the arguments about the complexity of well-written code vs. "spaghetti code" apply, but there is no way to show the causal relationship. So, the opponents of metrics might argue that, if a study of several companies shows that software was written by software engineers who always wore yellow shirts had statistically significantly fewer defects in their code, then companies would start requiring a dress code of yellow shirts! This illustration is, of course, hyperbole, but the point of correlation versus causality is made.

While in many cases these objections might be valid, like most things metrics can be either useful or harmful depending on how they are used (or abused).

7.5 FAULT TOLERANCE

7.5.1 WHAT ARE CHECKPOINTS?

Checkpoints are a way to increase fault tolerance. In this scheme, intermediate results are written to memory at fixed locations in code, called checkpoints, for diagnostic purposes (Figure 7.10). The data in these locations can be used for debugging purposes, during system verification, and during system operation verification.

If the checkpoints are used only during testing, then this code is known as a test probe. Test probes can introduce subtle timing errors, which are discussed later.

7.5.2 WHAT ARE RECOVERY BLOCKS?

Fault tolerance can be further increased by using checkpoints in conjunction with predetermined reset points in the software. These reset points mark recovery blocks in the software.

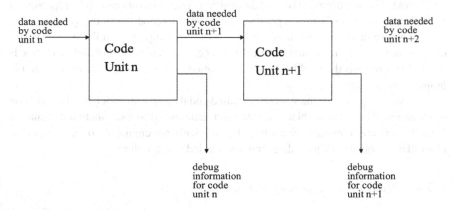

FIGURE 7.10 Checkpoint implementation

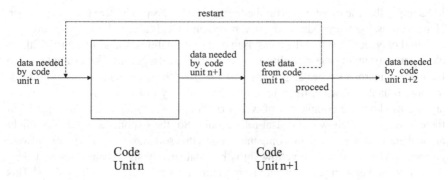

FIGURE 7.11   Recovery block implementation

At the end of each recovery block, the checkpoints are tested for "reasonableness." If the results are not reasonable, then processing resumes at the beginning of that recovery block or at some point in the previous one (see Figure 7.11).

The point, of course, is that some hardware device (or another process that is independent of the one in question) has provided faulty inputs to the block. By repeating the processing in the block, with presumably valid data, the error will not be repeated.

Each recovery block represents a redundant parallel process to the block being tested. Unfortunately, although this strategy increases system reliability, it can have a severe impact on performance because of the overhead added by the checkpoint and repetition of the processing in a block

7.5.3   WHAT ARE SOFTWARE BLACK BOXES?

The software black box is used in certain mission-critical systems. The objective of a software black box is to recreate the sequence of events that led to the software failure for the purpose of identifying the faulty code. The software black box recorder is essentially a checkpoint that stores behavioral data during program execution.

As procedural code is executed, control of execution passes from one code module to the next. This occurrence is called a transition. These transitions can be represented by an $N \times N$ matrix, where N represents the number of modules in the system.

When eaIh module i calls module j, the corresponding entry in the transition frequency matrix is incremented. From this process, a transition probability matrix is derived that records the likelihood that a transition will occur. This matrix is stored in nonvolatile memory.

Recovery begins after the system has failed and the software black box information is recovered. The software black box decoder generates possible functional scenarios from the transition probability matrix, allowing software engineers to reconstruct the most likely sequence of module execution that led to the failure.

7.5.4   WHAT IS N-VERSION PROGRAMMING?

Sometimes a system can enter a state whereby it is rendered ineffective or deadlocked. This situation can be due to an untested flow-of-control in the software for which there is no escape or resource contention. For life-support systems, avionics systems, power plant control systems, and other types of systems, the results can be catastrophic.

In order to reduce the likelihood of this sort of catastrophic error, redundant processors are added to the system. These processors are coded to the same specifications but by different programming teams. It is, therefore, highly unlikely that more than one of the systems can lock up under the same circumstances. This technique is called N-version programming.

While it may seem cost prohibitive to use this technique, there are certain cases where the approach makes sense. A prominent example is the master computer (also known as the General Purpose Computer or GPC) for the now extinct U.S. Space Shuttle. In that case, the GPC was redundantly hosted on five computers—four had identical hardware and software and the fifth had completely different software programmed by another development team and running on different hardware.

7.5.5 What Is Built-in-Test Software?

Built-in-test software is any hardware diagnostic software that is executed in real-time by the operating system. Built-in-test software can enhance fault tolerance by providing ongoing diagnostics of the underlying hardware.

Built-in-test software is especially important in embedded systems. For example, if the built-in-test software determines that a sensor is malfunctioning, then the software may be able to shut off the sensor and continue operation using backup hardware.

7.5.6 How Should Built-in-Test Software Include CPU Testing?

It is probably more important that the health of the CPU be checked than any other component of the system. A set of carefully constructed tests can be performed to test the health of the CPU circuitry.

7.5.7 Should Built-in-Test Software Test Memory?

All types of memory, including nonvolatile memory, can be corrupted via electrostatic discharge, power surging, vibration, or other means. This damage can manifest itself either as a permutation of data stored in memory cells or as permanent damage to the cell itself. Damage to the contents of memory is called soft error, whereas damage to the cell itself is called hard error. All memory should be tested both at initialization and during normal processing, if possible. There are various algorithms for efficiently testing memory, which is described in embedded software design books such as "Software Engineering for Image Processing Systems" (Laplante, 2004).

7.5.8 What about Testing Other Devices?

Devices such as sensors, motors, cameras, and the like need to be tested continually, or their own self-testing need to be monitored by the software and appropriate action taken if the device falls.

7.6 MAINTENANCE AND REUSABILITY

7.6.1 WHAT IS MEANT BY SOFTWARE MAINTENANCE?

Software maintenance is the "...correction of errors and implementation of modifications needed to allow an existing system to perform new tasks and to perform old ones under new conditions..." (Dvorak, 1994).

7.6.2 WHAT IS REVERSE ENGINEERING?

Generically, reverse engineering is the process of analyzing a subject system to identify its components. Reverse engineering is sometimes called renovation or reclamation. While there are negative connotations to reverse engineering, as in theft of a design, reverse engineering, in some form, is essential for the improvement of the design or implementation or for recovery of documentation in the case of a system that may have been acquired legitimately from a third party.

7.6.3 WHAT IS AN APPROPRIATE MODEL FOR SOFTWARE REENGINEERING?

Many embedded and engineering software systems are legacy systems; that is, they constitute the next generation of an existing system. Others borrow code from related systems. In any case, most systems need to have a long shelf life so that development costs can be recouped. Maintaining a system over a long period usually requires some form of reengineering; that is, a reverse flow through the software life cycle.

Figure 7.12 is a graphical representation of a reengineering process. The forward engineering flow represents a simple, three-phase waterfall model— requirements, design, and implementation.

Documentation recovery or redocumentation is the creation or revision of documentation from a given system, including requirements and design documentation

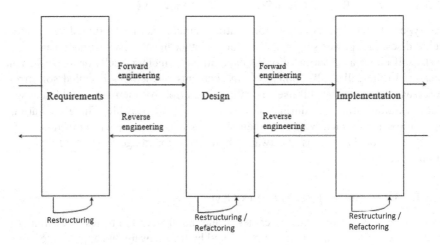

FIGURE 7.12 A reverse engineering process model

discovery. The need for redocumentation arises when there is poor or missing documentation for a number of reasons.

Design recovery is a subset of reverse engineering that recreates the design from code, existing documentation, personal insight, interviews with developers, and general knowledge. Again, the need for this arises in the case of poorly documented design, missing documentation, acquisition of a product from a company with inferior software engineering practices, and so on.

Restructuring is the transformation of one representation to another. In the case of code refactoring, the code is transformed so that the behavior is preserved. In design refactoring, the design is reengineered.

7.6.4 Since You Like Models So Much, Can You Give Me a Maintenance Process Model?

Of all of the phases, perhaps the maintenance model is the least understood. The maintenance phase generally consists of a series of reengineering processes to prolong the life of the system. There are three types of maintenance:

- Adaptive: Changes that result from external changes to which the system must respond
- Corrective: Changes that involve maintenance to correct errors
- Perfective: All other maintenance including enhancements, documentation changes, efficiency improvements, and so on

A widely adopted maintenance model illustrates the relationship between these various forms of maintenance (Figure 7.13).

The model starts with the generation of change requests by the customer, management, or the engineering team. An impact analysis is performed. This analysis involves studying the effect of the change on related components and systems and includes the development of a budget.

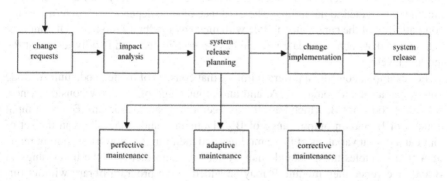

FIGURE 7.13 A maintenance process model (Adapted from Sommerville, 2015)

7.6.5   What Is System Release Planning?

System release planning involves the determination of whether the change is perfective, adaptive, or corrective. The nature of the change is crucial in determining whether the release needs to be made to all customers or specific customers, whether the release is going to be immediate or included in the next version, and so on. Finally, the change is implemented (invoking a mini-software life cycle process from concept to acceptance testing), followed by the official release of the new version.

7.6.6   What Is Software Reuse?

Pure software reuse is a highly sought prize in software engineering. It is clearly desirable to have a collection of mix-and-match, validated software components that could easily be pulled off-the-shelf for customized software applications. However, software reuse is virtually exploitation of hard-learned experience. Even if software modules are not being explicitly reused, the lessons learned from previous but similar software projects should be carried forward.

Most of the cost savings can then be expected by reusing domain-specific models. To reuse domain-specific logic, however, developers must clearly separate domain logic from that of the application. They must also clearly distinguish domain-independent logic.

Therefore, the best way to begin a program of software reuse is to start small and learn by doing. Try to identify several small software modules that are good candidates for reuse and focus on preparing these modules for that reuse.

7.6.7   Are There Special Techniques for Achieving Reuse in Procedural Languages?

One technique that has been used in building program libraries involves domain analysis. Domain analysis views software code as functions with an input domain and output range based on the range of their inputs.

The approach is as follows. In a set-theoretic way, define the input and output domains for each module to be added to the program library. Then, determine the input/output dependencies between each module in the library and any candidate module to be added to the library. The existence of such dependencies determines the compatibility of the candidate module with the existing library modules. Of course, it is assumed that each candidate module has been validated, and is fully tested, at the module level.

For example, consider a program library that consists of trusted code unit A. Code unit A has an input domain of A_I and an output range of A_O. Now consider a new candidate code unit B, which has already been unit tested. Code unit B has an input domain of B_I and an output range of B_O. "Domains" and "ranges" mean the set of input and output variables of these modules and their ranges. Now, if the output range of A (the variables that A could change) does not intersect with the input range of B and vice versa, then module B may be added to the program library without further interdependence or compatibility testing. If the input range of B and the output range of A overlap, then interdependencies and compatibility need to be tested before adding A to the library. Formally,

$$\text{If } A_0 \cap B_1 = \phi \text{ and } B_0 \cap A_1 = \phi \text{ then add A to the library} \qquad (7.6)$$
$$\text{Else test further before adding}$$

As additional modules or code are added to the library, interdependence testing must be completed for all modules in the library. For example, if A and B are trusted software in the library and module C is a candidate for the library, it must now be tested against A and B before adding it. Formally,

$$\text{If } A_0 \cap C_1 = \phi \text{ and } B_0 \cap C_1 = \phi \text{ and } C_0 \cap B_1 = \phi \text{ and}$$
$$B_0 \cap C_1 = \phi \text{ then add A to the library} \qquad (7.7)$$
$$\text{Else test further before adding}$$

It is easy to see that the level of effort grows rapidly as new code is added to the trusted program library.

7.6.8 Are There Special Techniques for Achieving Reuse in Object-Oriented Languages?

In object-oriented systems, the use of design patterns following the OOD principles previously described promotes reusability. For example, one way is to employ the protected variation (Parnas partitioning) principle by identifying those design aspects that are likely to change and building a stable interface around them. Design patterns can loosen the binding between program components enabling certain types of program evolution to occur with minimal changes to the program itself. However, to make good use of design patterns, the application's design process must undergo at least two iterations over the project life cycle.

7.6.9 When Is It Appropriate Not to Reuse Software?

It is sometimes desirable to plan not to reuse certain code. For example, throwaway prototypes are intentionally not to be reused. In other cases, it may not be desirable to try to reuse code that is of limited value. For example, a set of utilities intended for very specific hardware or that serves a very specialized function is probably not worth engineering for reuse when the hardware changes or becomes obsolete.

In any event, reuse of code that was not designed and coded for reuse can create many problems. For example, when a "quick-and-dirty" program becomes a widely used tool, it can present a maintenance nightmare.

7.6.10 What Is Pareto's Principle and How Does It Apply to Software Engineering?

Pareto was a late 19th and early 20th-century Italian mathematician and economist who was interested in the laws of chance. His observations can be applied in several ways to software reuse and engineering. For example, Pareto's principle might suggest that:

- 20% of the code contributes to 80% of the cost of software development.
- 20% of the code contributes 80% of the errors.
- 20% of the errors account for 80% of the cost to fix.
- 20% of the modules consume 80% of the execution time.

The percentages are, of course, arbitrary. But these observations provide insight into how to approach software reuse, testing, and effort planning. For example, it would be helpful to identify the 20% of software that is the most expensive to develop and plan to reuse that software. The other 80% that is relatively easy to develop might not be a prime candidate for reuse. Checkpoints and software black boxes can help to collect code unit execution frequency to identify the high-use code.

7.6.11 What Is the "Second System Effect"?

The Second System Effect first characterized by Brooks (1995) explains why software maintenance for legacy systems presents such challenges. This phenomenon is discussed in "The Mythical Man Month", a series of essays on software project management by Brooks published in the early 1980s. The essays are still relevant today. Brooks notes that "second systems" or the next generation of a delivered system tend to be over-engineered. That is, there is a tendency to carry over and refine techniques whose existence has been made obsolete by changes in basic system assumptions. Doing so tends to make these systems hard to maintain, unwieldy, and unreliable.

Consider, for example, a baggage handling system that was developed in the 1970s for hardware that is no longer available. In a second system, the underlying hardware may have been modernized. Hence, carrying over old design decisions can be disastrous. Embedded systems tend to be based on carry-over software, often originally written in Fortran, C, or assembly language and even BASIC. In some cases, C code is simply "objectified" by wrapping the C code in such a way that it can be compiled as C++ code.

Brooks suggests that the way to avoid this effect is to insist on a project leader who has had experience with at least two systems. In this recommendation, Brooks recognizes that software houses tend to assign new software engineers to maintain old legacy systems, while more senior engineers are assigned to new software development. While new projects may be more glamorous, younger engineers may not have the confidence or experience to challenge bad design decisions on a legacy system. Hence, it is probably better to have a combination of experience and youth assigned to both new and legacy system software development.

NOTE

1   There are other, equivalent, formulations; $C = e - n + p$ and $C = e - n + 2p$. The different forms arise from the transformation of an arbitrary directed graph to a strongly connected, directed graph obtained by adding one edge from the sink to the source node (Jorgensen & DeVris, 2021).

FURTHER READING

Basili, V.R., Caldiera, G., & Rombach, H.D. (1994). The goal question metric approach. Available at: www.cs.umd.edu/users/mvz/handouts/gqm.pdf

Boehm, B., *Software risk management: principles and practice. IEEE Software, 8*(1), 32–41.

Boehm, B., & Basili, V. (2005). Software defect reduction top-10 list. In *Foundations of Empirical Software Engineering.* Boehm, B., Rombach, H.D. & Zelkowitz, M., Eds., Springer, Secaucus, NJ, pp. 427–431.

Brooks, F.W. (1995). *The Mythical Man Month. 20th Anniversary Edition.* Addison-Wesley, Boston, MA.

Chidamber, S.R., & Kemerer, C.F. (1994). A metrics suite for object oriented design. *IEEE Trans. Software Eng., 20*(6), 476–493,.

Darcy, D., & Kemerer, C. (2005). OO metrics in practice. *IEEE Software, 22*(6), 17–19.

Dvorak, J. (1994). Conceptual entropy and its effect on class hierarchies, *Computer, 27*(6), 59–63.

Eickelmann, N. (2004). Measuring maturity goes beyond process. *IEEE Software,* 12–13.

Fischer, D., Augustine, D., & Pham, N.C. (Apr. 5, 2019). Was Boeing's Compensation Committee Sufficiently Independent in Judging the Business Risk of the 737 Max? [Online] . Available: https://ssrn.com/abstract=3370066

Godfrey, M., & Tu, Q. (September 2001). Growth, evolution and structural change in open-source software. *Proc. 2001 Intl. Workshop on Principles of Software Evolution (IWPSE-01), Vienna.*

Grady, R. (1992). *Practical Software Metrics for Project Management & Process Improvement.* Prentice Hall, Englewood Cliffs, NJ.

Hardgrave, W., & Armstrong, N.Y. (2005). Software process improvement: it's a journey not a destination. *Commun. ACM, 49*(11), 93–96.

IEEE Standard 829–2008 IEEE Standard for Software Test Documentation https://standards. ieee.org/ieee/829/3787/

Jones, C. (2007). *Estimating Software Costs. 2nd edition.* McGraw-Hill, New York.

Jorgensen, P.C. (2002) *Software Testing: A Craftsman's Approach. 2nd ed.* CRC Press, Boca Raton, FL.

Jorgensen, P.C., & DeVries, B. (2021). *Software Testing: A Craftsman's Approach.* 5th ed. Auerbach Publications, Boca Raton, FL.

Kandt, K. (2005). *Software Engineering Quality Practices.* Auerbach Publications, Boca Raton, FL.

Kassab, M., DeFranco, J.F., & Laplante, P.A. (2017). Software testing: The state of the practice. *IEEE Software, 34*(5), 46–52.

Kassab, M., Laplante, P., Defranco, J., Neto, V.V.G., & Destefanis, G. (2021). Exploring the Profiles of Software Testing Jobs in the United States. *IEEE Access, 9,* 68905–68916.

Koch, S. (2005). Evolution of open-source software systems — A large-scale investigation. *Proc. 1st Intl. Conf. OSS,* Scotto, M. & Succi, G. (Eds.), July 11–15, Genova, 148–153.

Kuhn, R., & Kacker, R. (2019). *An application of combinatorial methods for explainability in artificial intelligence and machine learning (draft).* National Institute of Standards and Technology.

Lakos, J. (1996). *Large-Scale C++ Software Design.* Addison-Wesley, Boston, MA.

Lange, C.F.J., Chaudron, M.R.V., & Muskens, J. (2006). UML software architecture and design description. *IEEE Software, 23*(2) 40–46.

Laplante, P.A. (2004). *Real-Time Systems Design and Analysis.* Third Edition. John Wiley & Sons/IEEE Press.

Laplante, P.A. (2004). *Software Engineering for Image Processing Systems.* CRC Press, Boca Raton, FL.

Machiavelli, N. (1513). *The Prince.*

Martin, R.C. (2002). *Agile Software Development: Principles, Patterns, and Practices.* Prentice Hall, Englewood Cliffs, NJ.

McCabe, T. (1976). A software complexity measure. *IEEE Trans. Software Eng., SE-2,* 308–320.

Morozoff, E. (January/February 2010). Using a Line of Code Metric to Understand Software Rework. *IEEE Software*, pp. 72–77.

Munson, J. (2003). *Software Engineering Measurement.* Auerbach Publications, Boca Raton, FL.

Naik, K. & Tripathy, P. (2008. *Software testing and quality assurance: theory and practice.* John Wiley & Sons.

Nakakoji, K., Yamamoto, Y., Nishinaka, Y., Kishida, K., & Ye, Y. (May 2002). Evolution patterns of open-source software systems and communities. *Proc. Intl. Workshop on Principles of Software Evolution,* ACM Press, Orlando, FL.

NIST, *Planning Report 02-3, The Economic Impact of Inadequate Infrastructure for Software.* www.nist.gov/director/prog-ofc/report02-3.pdf.

Paulk, M.C., Weber, C.V., Curtis, B., & Chrissis, M.B. (1995). *The Capability Maturity Model, Guidelines for Improving the Software Process.* Addison-Wesley, Boston, MA.

Pfleeger, S.L. (1992). Measuring software reliability. *IEEE Spectrum,* 55–60.

Pressman, R.S., & Maxim, B. (2019). *Software Engineering: A Practitioner's Approach.* 9th ed. McGraw-Hill, New York.

Raymond, E.S. (2001). *The Cathedral and the Bazaar. Musings on Linux and Open-source by an Accidental Revolutionary.* O'Reilly Media, Cambridge, MA.

Scacchi, W. (2004). Understanding open-source software evolution. in *Software Evolution.* Madhavji, N.H., Lehman, M.M., Ramil, J.F., & Perry, D. (Eds.), John Wiley & Sons, New York.

Sharp, J. et al. (2005). Tensions around the adoption and evolution of software quality management systems: a discourse analytic approach. *Journal of Human-Computer Studies, 61,* 219–236.

Sommerville, I. (2015). *Software Engineering.* 10th ed. Pearson, London, UK.

Stelzer, D., & Mellis, W. (1998). Success factors of organizational change in software process improvement. *Software Process — Improvement and Practice, 4,* 227–250.

Voas, J.M., & Agresti, W.W. (2004). Software quality from a behavioral perspective. *IT Pro,* 6(4), 46–50.

Voas, J.M., & McGraw, G. (1998)., *Software Fault Injection: Inoculating Programs Against Errors.* John Wiley & Sons, New York.

West, M. (2004). *Real Process Improvement Using CMMI.* Auerbach Publications, Boca Raton, FL.

Wong, W.E., Li, X., & Laplante, P.A. (2017). Be more familiar with our enemies and pave the way forward: A review of the roles bugs played in software failures. *Journal of Systems and Software 133*, pp. 68–94, Nov.

Zubrow, D. (2003). Current trends in the adoption of the CMMI® product suite, *compsac. 27th Annu. Intl. Comp. Software Appl. Conf.*, 126–129.

8 Managing Software Projects and Software Engineers

OUTLINE

- Software engineers are people too
- Project management basics
- Antipatterns in organizations management
- Tracking and reporting progress
- Software cost estimation
- Project cost justification
- Risk management

8.1 INTRODUCTION

For some reason, a stereotype exists that engineers lack people skills. In our experience, this unfair perception of engineers is prevalent in those who have little or no education in the sciences or mathematics. These folks apparently think that engineers, mathematicians, and scientists lack soft skills because they have mastered the more analytical disciplines.

Of course, some engineers or scientists have unpleasant personalities. But there are non-engineers with rotten demeanors, too. We have never seen a study by any behavioral scientists demonstrating that lousy personalities are found at a higher frequency in engineers than in any other profession. The point here is that it doesn't matter who you are or where you came from—if you are a manager of any type, even a software project manager, you need the right attitude, education, and experience to be a good software project manager. In this chapter, we will examine the people aspects of software project management as well as the more technical issues of cost determination, cost justification, progress tracking and reporting, and risk management.

While some of the topics are unique to software projects, most of the ideas presented in this chapter can be applied across a wide spectrum of technical and nontechnical projects. Note that some of this chapter is adapted and reused, with permission, from other books by the authors, notably Laplante & Kassab (2022), Neill, Laplante, & DeFranco (2012), and Laplante (2003).

DOI: 10.1201/9781003218647-9

8.2   SOFTWARE ENGINEERS ARE PEOPLE TOO

8.2.1   WHAT PERSONNEL MANAGEMENT SKILLS DOES THE SOFTWARE PROJECT MANAGER NEED?

A project manager needs to have an appropriate set of people skills and relevant technical skills. The people skills include team building, negotiation techniques, understanding of psychology and group dynamics, good motivational skills, and excellent communication skills (especially listening). Most of the people's skills involve self-improvement, and it is beyond the scope of this book to delve too deeply into that. There are many good books on the self-improvement aspects of people management, some of which can be found in the Further Reading section of this chapter. We will discuss some of the team-building aspects of people management, however.

8.2.2   BUT WHAT'S THE BIG DEAL WITH "PEOPLE ISSUES"?

It is well-known that the success of a project is directly related to the quality of talent employed and, more importantly, the manner in which the talent is deployed on the project. But too frequently project managers[1] view themselves as technical managers only, forgetting that human nature enters into technical situations.

Moreover, the special challenges of developing software are imposed on top of the already daunting challenge of managing human teams. Some people might consider the aspect of human resource management insignificant if the project team has enough technical skill. This is generally not true.

8.2.3   HOW DOES TEAM CHEMISTRY AFFECT SOFTWARE PROJECTS?

The key problem in most cases is that the chemistry of the team makes it impossible for the manager to overcome other constraints, such as technical, time, and budget, even with good people. Table 8.1 illustrates the four possible cases of good/bad management and good/bad team chemistry.

In the case where both management and chemistry are good, the likelihood of project success (which itself, must be carefully defined) is high. In the case of bad management, success is unlikely even with good team chemistry because bad management will eventually erode morale. Even when team chemistry is bad, good management can possibly lead to success.

TABLE 8.1
Four Possible Combinations of Good/Bad Management and Good/Bad Team Chemistry

	Good team chemistry	Bad team chemistry
Good management	Likely success	Possible success
Bad management	Unlikely Success	Unlikely Success

8.2.4   Why Is Team Chemistry So Hard to Manage?

One reason is that the number of working relationships grows as a polynomial function of n, the number of people on the team. This might be whimsically referred to as the "n-body problem." In fact, it can easily be shown that for n people on a team, there are $\dfrac{n(n-1)}{2}$ possible working relationships, any of which can sour. Furthermore, a working relationship is not transitive. So, for example, Roger may work well with Mary and Mary with Sue, but Roger and Sue may not work well together. Finally, complicating these interactions are intercultural differences and outsourcing of project components. All of these aspects must be considered when building and managing teams, planning projects, and dealing with difficult personnel situations. "Too many cooks spoil the broth" or, to paraphrase Brooks (1995), "adding manpower to a late software project makes it later."

8.2.5   Management Styles

8.2.5.1   What Are Some Styles for Leading Teams?

There are almost as many management styles as there are people. But, traditionally, a small collection of paradigms can be used to describe the management style of an individual or organization more or less. Understanding these basic approaches can help understand the motivations of customers, supervisors, and subordinates.

8.2.5.2   What Is Theory X?

Theory X, perhaps the oldest management style, is closely related to the hierarchical, command-and-control model used by military organizations. Accordingly, this approach is necessary because most people inherently dislike work and will avoid it if they can. Hence, managers should coerce, control, direct, and threaten their workers in order to get the most out of them. A typical statement by a "Theory X" manager is "people only do what you audit."

8.2.5.3   What Is Theory Y?

As opposed to Theory X, Theory Y holds that work is a natural and desirable activity. Hence, external control and threats are not needed to guide the organization. In fact, the level of commitment is based on the clarity and desirability of the goals set for the group. Theory Y posits that most individuals actually seek responsibility and do not shirk it, as Theory X proposes.

A Theory Y manager simply needs to provide the resources, articulate the goals, and leave the team alone. This approach doesn't always work, of course, because some individuals do need more supervision than others.

8.2.5.4   What Is Theory Z?

Theory Z is based on the philosophy that employees will stay for life with a single employer when there is strong bonding to the corporation and subordination of individual identity to that of the company. Theory Z organizations have implicit, not explicit, control mechanisms such as peer and group pressure. The norms of the particular corporate culture also provide additional implicit controls. Japanese companies are well-known for their collective decision making and responsibility at all levels.

Theory Z management emphasizes a high degree of cross-functionality for all of its workers. Specialization is discouraged. Most top Japanese managers have worked in all aspects of their business from the production floor to sales and marketing. This is also true within functional groups. For example, assemblers will be cross-trained to operate any machine on the assembly floor. Theory Z employers are notoriously slow in giving promotions, and most Japanese CEOs are over age 50.

The purpose of this litany of alphabetic management styles is not to promote one over another; in fact, we don't recommend adopting any of these naïvely. But many individual team members and managers will exhibit some behaviors from one of the above styles, and it is helpful to know what motivates them. Finally, certain individuals may prefer to be managed as a Theory X or Theory Y type (Theory Z is less likely in this case), and it is good to be able to recognize the signs. Moreover, some companies might be implicitly based on one style or another.

8.2.5.5   What Is Theory W?

Theory W is a software project management paradigm developed by Boehm (1989), which focuses on the following for each project:

- Establishing a set of win–win preconditions
- Structuring a win–win software process
- Structuring a win–win software product.

8.2.5.6   What Does It Mean to Establish a Set of Win–Win Preconditions?

This means recognizing that the best working relationships are those in which everyone "wins." Zero-sum, win–lose, or lose–win situations can leave one or both parties bitter.

Win–win solutions can be sought as follows. First, recognize that everyone wants to win. Then, understand what constitutes a winning situation for each individual. Money, power, and recognition contribute to winning conditions for most people, but there are other, more subtle, conditions such as job satisfaction, a feeling of belonging, and moral fulfillment.

Next, establish reasonable expectations. The importance of setting reasonable and mutually fulfilling expectations in every aspect of human relations can't be overemphasized. Then, ensure that task assignments match the win conditions.

Finally, provide an environment that supports the fulfillment of the win conditions. This can take a variety of forms but might include such things as financial incentives, group activities, and communication sessions to head off problems.

8.2.5.7   What Does It Mean to Structure a Win–Win Software Process?

It means setting up a software process that will lead to success. This includes establishing a realistic process plan based on some standard methodology. This methodology may be internal and company-wide, or off-the-shelf.

It is also important to use the project/management plan to control the project. Too often, managers develop a project plan to sell the job to senior management or the customer, and then throw the plan away once it is approved. Therefore, be sure to use and maintain the project plan throughout the life of the project.

Project managers also need to monitor the risks that have been described as they can lead to win–lose or lose–lose situations. Thus, risks should be identified and eliminated at the earliest opportunity.

Keeping people involved is essential. It helps team members feel a part of the project and improves communications. Besides, listening to team members can reveal great ideas.

8.2.5.8 What Does Structuring a Win–Win Software Product Mean?

This refers to the process of specification writing by matching the users' and maintainers' win conditions. This process also requires careful and honest expectation setting.

8.2.5.9 What Is Principle-Centered Leadership?

All of the management approaches discussed thus far focus on organizational frameworks for management. Principle-Centered Leadership focuses on the behavior of the manager as an agent for change (Covey, 1991). Some management theorists hold that motivating team members by example and leadership, and not through the hierarchical application of authority, is much more effective (manage things, but lead people). A key concept in Principle-Centered Leadership is that the best managers are leaders and that the only way to affect change is by the managers changing themselves first.

Principle-Centered Leadership recognizes that principles are more important than values. Values are society-based and can change over time and differ from culture to culture. Principles are more universal, more lasting. Think of some of the old principles like the "Golden Rule." That is, treat others as you would like to be treated. These kinds of principles are timeless and transcend cultures. In fact, there is a great deal of similarity in Principle-Centered Leadership and Theory W, with Principle-Centered Leadership being much more generic.

8.2.5.10 What Is Management by Sight?

Also known as management by walking around, this is not really a full-bodied management approach, but rather a sub-strategy for the approaches already discussed. This approach is people-oriented because it requires the manager to be very visible and to interact with staff. Interacting with staff at all levels is a good way for managers to collect important information about the project and the people in their care.

Management by sight is obvious. The manager uses observation and visibility to provide leadership, to monitor the situation, and even to control when necessary. In general, it is advisable to incorporate this strategy into any management approach.

8.2.5.11   What Is Management by Objectives?

Management by objectives (MBO) is another sub-strategy that can be used in conjunction with any other management approach. MBO involves managers and subordinates jointly setting carefully structured objectives with measurable outcomes and rewards.

Coupled with periodic reviews to measure progress, MBO has the effect of positive reinforcement of desired performance. For example, a manager may agree with the team member responsible for writing a section of the SDD to complete the task by a certain date (provisos can be made for various inevitable distractions that will appear). In return, time off might be granted for meeting the goal: more time off for early completion. The scenario becomes somewhat more complex when the other ten things for which the team member is responsible are factored in (e.g., producing other reports, attending meetings, and working on another project simultaneously). Process tracking tools are very helpful in this case. But the real keys to MBO are setting reasonably aggressive goals and having a clear means of measuring success.

8.2.5.12   What Is Pascal's Wager Theory to Manage Expectations?

Mathematician Blaise Pascal (1623–1662) is well-known for various achievements including Pascal's Triangle (a convenient way to find binomial coefficients) and work in probability. It was his work in probability theory that led to the notion of expected value, and he used such an approach later in life when he became more interested in religion than mathematics. It was during his monastic living that he developed a theory that suggested that it was advisable to live a virtuous life, whether or not one believed in a supreme being. His approach, using expected value theory, is now called Pascal's Wager, and it goes like this.

Imagine an individual is having a trial of faith and is unsure if they believe in this supreme being (let's call this being God for argument's sake) or not. Pascal suggests that it is valuable to consider the consequences of living virtuously, in the face of the eventual realization that such a God exists (or not). To see this, consider Table 8.2.

Assuming that it is equally probable that God exists or not (this is a big assumption), we see that the expected outcome (consequence) of living virtuously is half of the paradise while the expected outcome of living without virtue is half of the damnation. Therefore, it is in a person's best interests to live virtuously.

TABLE 8.2
Pascal's Wager Consequence Matrix

	God exists	God does not exist
Live virtuously	Achieve paradise	Null
Do not live virtuously	Achieve damnation	Null

8.2.5.13   What Does Pascal's Wager Have to Do with Expectation Setting in Project Development?

Stakeholders will hedge their bets—sometimes withholding information or offering up inferior information because they are playing the odds involving various organizational issues. For example, does a stakeholder wish to request a feature that they believe no one else wants and for which they might be ridiculed? From a game theory standpoint, it is safer to withhold their opinion. To see this, consider the modified Pascal's Wager outcome matrix in Table 8.3.

If the stakeholder decides to speak out about a particular feature (or in opposition to a particular feature), assuming equi-likely probability that the group will agree or disagree, the consequence matrix shows that they can expect to get some praise if the group agrees or some ridicule if the group disagrees. It is also well-known that, in decision making, individuals will tend to make decisions that avoid loss over those that have the potential for gain—most individuals are risk-averse. Furthermore, recall the possible cultural effects for stakeholders who originate from countries with high power distances and masculinity indices. The foregoing analysis also assumes that the probabilities of agreement and disagreement are the same—the expected consequences are much worse if the stakeholder believes there is a strong chance that their colleagues will disagree. Of course, later in the process, the feature might suddenly be discovered by others to be important. Now it is very late in the game, however, and adding this feature is costly. Had the stakeholder only spoken up in the beginning, in a safe environment for discussion, a great deal of cost and trouble could have been avoided.

8.2.5.14   Any Other Comments on Pascal's Wager, Expectation, and Risk Management?

One of the authors once worked for a boss—we'll call him Bob—who greeted all new employees with a welcome lunch. At that lunch, he would declare "I am a 'no surprises' kind of guy. You don't surprise me, and I won't surprise you. So, if there is ever anything on your mind or any problem brewing, I want you to bring that to my attention right away." This sentiment sounded great. However, each time the author or anyone else would bring bad news to Bob, whether the messenger was responsible for the situation or not, Bob would get very angry and berate the hapless do-gooder. After a while, potential messengers would forego bringing information to Bob. The decision was purely game-theoretic—if you had bad news and you brought it to Bob,

---

TABLE 8.3
Modified Pascal's Wager Consequence Matrix

	Group agrees	Group disagrees
Speak out	Get praise	Get ridiculed
Remain silent	Nothing happens	Nothing happens

he would yell at you. If he didn't find out about it (as often happened because no one else would tell him, either), then you escaped his rampage. If he somehow found out about the bad news, you might get yelled at—but he might yell at the first person in his sight, not you, even if you were the party responsible for the problem. So, it made sense (and it was rigorously sound via game theory) to stay quiet.

It was rather ironic that "no surprise Bob" was always surprised because everyone was afraid to tell him anything. The lesson here is that, if you shoot the messenger, people will begin to realize the consequences of bringing you information, and you will soon be deprived of that information. Actively seeking and inviting requirements information throughout the project lifecycle is an essential aspect of requirements management.

8.2.6 DEALING WITH PROBLEMS

8.2.6.1 How Do I Deal with Difficult People?

Whether they are subordinates, peers, or superiors, dealing with difficult people is always a challenge. The first thing to do is to avoid forming an opinion too soon. Never attribute some behavior to malice when a misunderstanding could be the reason. Almost without exception, taking the time to investigate an issue and to think about it calmly is superior to reacting spontaneously or emotionally.

Whatever management style is employed, the manager should make sure that the focus is on issues and not people. Managers should avoid the use of accusatory language such as telling someone they are incompetent. The manager should focus, instead, on their feelings about the situation.

Make sure that all sides of the story are listened to when arbitrating a dispute before forming a plan of resolution. It is often said that there are three sides to an issue, the sides of the two opponents and the truth, which is somewhere in between. While this is a cliché, there is much truth to it. The manager should always work to set or clarify expectations. Management failures, parental failures, marital failures, and the like, are generally caused by a lack of clear expectations. The manager should set expectations early in the process, making sure that everyone understands them. They should continue to monitor the expectations and refine them if necessary.

Good team chemistry can be fostered through mentoring and most of the best managers fit the description of a mentor. The behaviors already described are generally those of someone who has a mentoring personality. Finally, the manager should be an optimist. No one chooses to fail. The manager should always give people the benefit of the doubt and work with them.

8.2.6.2 Is That It? Can't You Give Me a Playbook for Handling Difficult Situations?

Team management is a complex issue and there are many books on the subject and a great deal of variation in how to deal with challenging situations. Table 8.4 is a summary of various sources of conflict and a list of things to try to deal with those conflicts.

TABLE 8.4
Various Sources of Conflict and Suggestions on How to Manage Them Based on Analysis of Experience Reports from 56 Companies

	Sources of Conflict	Managing Conflict
Process	Scarce resource of time	• Employ time management.
		• Plan for schedule overruns
		• Manage effect of schedule changes
		• Learn from project experience
	User vs. technical requirements	• Identify common goals
		• Align individual goals with process metrics
		• Value team more than individual success
People	Disagreement	• Apply team-building principles
		• Train in conflict resolution
		• Sponsor group activities
		• Support informal social contact
	Personalization of code	• Understanding of one another's point of view
Organization	Power and politics	• Structure for success
		• Co-locate teams
		• Integrate development/testing functions
		• Instill ownership
	Management behavior	• Get leadership involved
		• Create a collaborative atmosphere
		• Model effective conflict management

8.2.6.3 How Do I Manage Divergent Agendas in a Software Project?

Each stakeholder has a different agenda. For example, business owners seek ways to get their money's worth from projects. Business partners want explicit requirements because they are like a contract. Senior management expects more financial gain from projects than can be realized. And systems and software developers like uncertainty because it gives them the freedom to innovate solutions. Project managers may use the requirements to protect them from false accusations of underperformance in the delivered product.

One way to understand why the different agendas might exist—even among persons within the same stakeholder group—is the Rashomon effect. *Rashomon* is a highly revered 1950 Japanese film directed by Akira Kurosawa. The main plot involves the recounting of the murder of a samurai from the perspective of four witnesses to that event—the samurai, his wife, a bandit, and a woodcutter—each of whom has a hidden agenda, and tells a contradicting accounting of the event. Stated succinctly "your understanding of an event is influenced by many factors, such as your point of view and your interests in the outcome of the event" (Lawrence, 1996). The smart requirements manager seeks to manage these agendas by asking the right questions upfront. Andriole (1998) suggests the following questions are appropriate:

1. What is the project request?
   a. Who wants it?
   b. Is it discretionary or non-discretionary?
2. What is the project's purpose?
   a. If completed, what impact will the new or enhanced system have on organizational performance?
   b. On profitability?
   c. On product development?
   d. On customer retention and customer service?
3. What are the functional requirements?
   a. What are the specific things the system should do to satisfy the purposeful requirements?
   b. How should they be ranked?
   c. What are the implementation risks?
4. What are the non-functional requirements, like security, usability, and interoperability?
   a. How should they be ranked?
   b. What are the implementation risks?
   c. How do you trade off functional and non-functional requirements?
5. Do we understand the project well enough to prototype its functionality?
6. If the prototype is acceptable, will everyone sign off on the prioritized functionality and non-functionality to be delivered, on the initial cost and schedule estimates, on the estimates' inherent uncertainty, on the project's scope, and on the management of additional requirements?

Andriole asserts that, by asking these questions upfront, hidden agendas can be uncovered and differences resolved. At the very least, important issues will be raised upfront and not much later in the process.

8.2.6.4   What Are the Approaches for Consensus Building among Stakeholders?

There are numerous approaches to consensus building. Existing trade-off techniques are categorized and sorted into three categories: experience-based, model-based, and mathematically based (Berander et al., 2005).

Experience-based models rely purely on experience for supplying the required information to perform the trade-off analysis. They are commonly used for cost and effort estimation, but rarely in literature, as they are ad hoc in nature, and the execution is up to the person performing the trade-off. These models provide quick analysis, but they are not repeatable, and they do not provide any figures presenting the trade-off.

Model-based techniques; on the other hand, rely on constructing, for example, a graphical model for illustrating and concretizing the relations between trade-off entities, thus facilitating the trade-off. By applying a model-based trade-off approach, in comparison to an experience-based approach, it is possible to structure and communicate the knowledge. Dealing with trade-offs concerning quality requirements is popularly treated through the model-based technique.

Mathematically based trade-off techniques, on the other hand, rely on formalization for constructing and representing the trade-off, thus making it possible to feed the mathematical construct with appropriate values and receive the best solution (either maximization or minimization or optimal with regard to certain criteria). Mathematically based trade-off techniques are widely used in management for trade-off decision support. Estimation and calculations regarding break-even, optimum production volume, and so on all use mathematical techniques. Examples of these techniques include: the Analytical Hierarchy Process (AHP) technique, Wideband Delphi technique, and reliability growth methods. Mathematically based trade-off techniques can handle large amounts of variables and come up with results that are generally more accurate than common sense. It also enables repeatable and structured analysis. If a measurement program is in place collecting metrics, and if mathematically based trade-off techniques are used as a way to estimate issues, the work can be replicated over several releases tweaking the data and choice of the technique to correspond with needs. Using the same types of techniques over an extended period of time can give an organization consistency and overview.

8.2.6.5 What Is Analytical Hierarchy Process?

The analytical hierarchy process (AHP) is a technique for modeling complex and multi-criteria problems and solving them using a pairwise comparison process. Based on mathematics and psychology, it was developed by Thomas L. Saaty in the 1970s and has been extensively studied and refined since then. AHP was refined through its application to a wide variety of decision areas, including transport planning, product portfolio selection, benchmarking and resource allocation, and energy rationing.

Simply described, AHP breaks down complex and unstructured problems into a hierarchy of factors. A super-factor may include sub-factors. By pairwise comparison of the factors in the lowest level, we can obtain a prior order of factors under a certain decision criterion. The prior order of super-factors can be deduced from the prior order of sub-factors according to the hierarchy relations.

The AHP process starts with a detailed definition of the problem: goals, all relevant factors, and alternative actions are identified. The identified elements are then structured into a hierarchy of levels where goals are put at the highest level and alternative actions are put at the lowest level.

Usually, an AHP hierarchy has at least three levels: the goal level, the criteria level, and the level of the alternative (see Figure 8.1). This hierarchy highlights relevant factors of the problem and their relationships to each other and to the system as a whole.

Once the hierarchy is built, involved stakeholders (i.e., decision-makers) judge and specify the importance of the elements of the hierarchy. To establish the importance of elements of the problem, a pairwise comparison process is used. This process starts at the top of the hierarchy by selecting an element (e.g., a goal) and

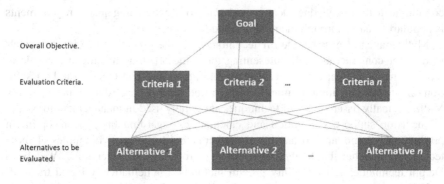

FIGURE 8.1 The AHP hierarchy

TABLE 8.5
Pairwise Comparison Scale for AHP (Saaty, 1988)

Intensity of Judgement	Numerical Rating
Extreme Importance	9
Very Strong Importance	7
Strong Importance	5
Moderate Importance	3
Equal Importance	1
For compromise between the above values	2, 4, 6, and 8

---

then the elements of the level immediately below are compared in pairs against the selected element.

A pairwise matrix is built for each element of the problem; this matrix reflects the relative importance of elements of a given level with respect to a property of the next higher level. Saaty (1988) proposed the scale [1,3,5,…,9] to rate the relative importance of one criterion over another (See Table 8.5).

Based on experience, a scale of 9 units is reasonable for humans to discriminate between preferences for two items. One important advantage of using the AHP technique is that it can measure the degree to which a manager's judgments are consistent. In the real world, some inconsistency is acceptable and even natural. For example, in a sporting contest, if team A usually beats team B, and if team B usually beats team C, this does not imply that team A usually beats team C. The slight inconsistency may result because of the way the teams match up overall. The point is to make sure that inconsistency remains within some reasonable limits. If it exceeds a specific limit, some revision of judgments may be required. AHP technique provides a method to compute the consistency of the pairwise comparisons.

Some examples of using the AHP in requirements prioritizations include the integrated approach of AHP and NFRs framework (Kassab, 2013; Kassab & Ergin, 2015). In addition, Kassab (2014) has used the AHP for effort estimation at requirements time.

8.2.6.6 What Is Wideband Delphi Technique?

One of the most celebrated consensus building techniques in systems engineering is the Wideband Delphi technique. Developed in the 1970s, but popularized by Barry Boehm in the early 1980s, Wideband Delphi is usually associated with the selection and prioritization of alternatives. These alternatives are usually posed in the form of a question:

> For the following alternative, rate your preference according to the following scale (5=most desired, 4=desired, 3=ambivalent, 2=not desired, 1=lease desired).

The list of alternatives and associated scale is presented to a panel of experts (and, in the case of requirements ranking, stakeholders) who rank these requirements silently and, sometimes, anonymously. The ranking scale can have any number of levels. The collected list of rankings is then presented back to the group by a coordinator. Usually, there is significant disagreement in the rankings. A discussion is conducted, and experts are asked to justify their differences in opinion. After discussion, the independent ranking process is repeated (Figure 8.2).

With each iteration, individual customer rankings should start to converge, and the process continues until a satisfactory level of convergence is achieved.

A more structured form of the process is:

1. The coordinator presents each expert with a specification and an estimation form.
2. The coordinator calls for a group meeting in which the experts discuss estimation issues with the coordinator and each other.
3. Experts fill out forms anonymously.
4. The coordinator prepares and distributes a summary of the estimates.

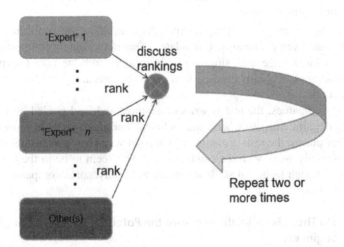

FIGURE 8.2 The Wideband Delphi process

5. The coordinator calls for a group meeting, specifically focusing on having the experts discuss points where their estimates vary widely.
6. Experts fill out forms, again anonymously, and steps 4 to 6 are iterated for as many rounds as appropriate.

There will never be a unanimous agreement in the Wideband Delphi process, but at least everyone involved will feel that their opinion has been considered. Wideband Delphi is a kind of win–win negotiating and can be used for other types of decision-making.

8.2.7   HIRING SOFTWARE ENGINEERING PERSONNEL

8.2.7.1   We Need to Hire More Software People. How Should I Approach This Task?

First, really consider whether you need another person on the team. Remember Brooks' admonition that adding more people to an already late project will make it later. If you need to hire more staff, say, because the scope of work has expanded or you have more work than the current staff can handle, remember Boehm's (1989) five staffing principles:

- The principle of top talent: Use better and fewer people.
- The principle of job matching: Fit the tasks to the skills and motivation of the people available (remember this when we talk about outsourcing/ offshoring).
- The principle of career progression: An organization does best in the long run by helping its people to self-actualize.
- The principle of team balance: Select people who will complement and harmonize with one another.
- The principle of phaseout: Keeping a misfit on a team doesn't benefit anyone.

8.2.7.2   I Want to Select the Right People, but How Is It Done in the Software Industry?

Companies use a variety of testing instruments to evaluate job applicants for skills, knowledge, and even personality. Candidate exams range widely from programming tests, general knowledge tests, situation analysis, and "trivia" tests that purport to test critical thinking skills. Many times, domain-specific tests are given. Some companies don't test at all.

In some companies, the test is written, administered, and graded by the Human Resources department. Elsewhere, the technical line managers administer the test. In still other places, the tests are organized and delivered by "peer" staff—those who would potentially work with the candidate and who seem to be in the best position to determine which technical skills are relevant. Finally, some companies outsource this kind of testing.

8.2.7.3   Do These Tests Really Measure the Potential Success of the Software Engineer?

There seems to be no consensus, even among those who study human performance testing, that these tests strongly correlate with new employee success.

One thing a company could do to test the efficacy of a knowledge/skills exam is to give it to a group of current employees who are known to be successful, and to another group who are less successful. Correlating the results could lead to a set of questions that "good" employees are more likely to get right (or wrong). Alternatively, a company could survey those employees who were fired for their poor skills to identify any common trends. Clearly though, while such studies might be interesting, they would be nearly impossible to conduct and, in any case, it is unclear if these tests measure what a hiring manager really wants to know.

Perhaps skill or knowledge testing is not what is needed. In many cases, a failure in attitude leads to performance shortfalls. Perhaps then some sort of assessment of the potential to get along with others might be needed. Some organizations will measure a candidate's compatibility with the rest of the team by testing "emotional intelligence," or personality. The idea is to establish the "style" of existing team members based on their personalities, then look to add someone whose style is compatible.

In any case, it is questionable as to whether these kinds of tests really do lead to better hires. Unfortunately, it is also hard to gauge a person's fit with the team based on a series of interviews with team members and the obligatory group lunch interview. Anyone can role-play for a few hours in order to get hired, and most interviewers are not that well-trained to see through an act.

8.2.7.4 You Don't Seem to Like These Tests. How Do I Assess the Potential of a Candidate Besides Checking References?

If you must use an assessment of "intelligence," college grades or graduate record examination scores may be used. If the real goal is to determine programming prowess, checking grades in programming courses might be helpful. Better still, the manager can ask the candidate to bring in a sample of some code they developed and discuss it. If the code sample is too trivial or the candidate struggles to explain it, it is likely they didn't write it or understand it that well.

Other skills can be tested in this manner as well. For example, if the job entails writing software specifications, design, or test plans, ask the candidate to provide a writing sample. It is possible the candidate does not have samples from their current or past job for proprietary or security reasons, but they should still be able to talk about what they did without notes. If they can recount the project in detail, then they probably know what they are talking about.

8.2.7.5 How Do You Measure the Candidate's Compatibility with Existing Team Members?

Many companies use personality tests, and in some cases, this might work. But there is no magic test here. The manager has to do their homework. They must spend time with the candidate in a variety of settings including having lunch (you can learn a lot about a person from their table manners, for example). Make sure that whoever will be working closely with this person is on the hiring/interview team and gets to meet the candidate. Also, make sure that everyone has been trained on what to look for during an interview. Then get together and compare notes right after the candidate leaves.

8.2.7.6 Are There Studies on the Most In-Demand Skills for Various Software Engineering Positions?

Yes. For example:

- A study by Wang, Cui, Daneva, & Kassab (2018) on understanding what industry wants from requirements engineers
- A study by Kassab, Laplante, DeFranco, Neto, & Destefanis (2021) on exploring the profiles of software testing jobs
- A study by Kassab, Destefanis, DeFranco, & Pranav (2021) on the profile of blockchain software engineers.

8.2.7.7 How Should I Reference-Check a Potential Hire?

Most people don't do a good job of checking references. Here are some simple guidelines for reference checking:

- Ask legal questions. Many questions cannot be asked, and human resources or legal advisors should be consulted before writing the interview questions.
- Evaluate the references the candidate has given you. If they have difficulty providing references, if none is a direct supervisor (or subordinate), then this may indicate a problem. If the referee barely knows the candidate or simply worked in the same building, then regardless of their opinions, they cannot be weighed heavily upon.
- Be sure you check at least three references. It is harder to hide any problems this way. Be sure to talk to supervisors, peers, and subordinates. A team member has to be able to lead and be led.
- Ask a "hidden" reference. Try to track down someone who would know the candidate, but who is not on their reference list. To avoid potential problems with their current employer, go back to one of their previous employers and track down someone who knew them well there. This strategy is very effective to find out if the candidate has any hidden issues.
- Take good notes and ask follow-up questions. Many referees are reluctant to say bad things about people even if they do not believe the person is a strong candidate. By listening carefully to what references do and do not say, the real message will come through.
- Be sure to ask a broad range of questions, and questions that encourage elaboration (as opposed to "yes" or "no" questions). For example, some of the following might be helpful:
  - Describe the candidate's technical skills.
  - Describe a difficult situation the candidate encountered and how they dealt with it.
  - Describe the candidate's interpersonal skills.
  - Why do you think the candidate is leaving the company?
  - Describe the kind of work environment in which the candidate would thrive.
  - Describe the current work environment in which the candidate works.

- Describe the contributions of the candidate in the capacities with which you are familiar.
- Describe the kind of manager you think would be best for the candidate.

In summary, it is crucial that the hiring manager and hiring/interview team learn the art of interviewing and background checking. This is more likely to lead to the right fit than a series of tests. Relying on "trivia tests" is risky at best and you might lose a good employee in the process.

8.2.7.8   Should I Check Social Network Posts of the Candidate?

It is important to check the social network posts of candidates relatively early in the hiring process, for example, when a shortlist of potential candidates has been made and before a final interview.

If there are any concerns about certain posts that may be offensive, inflammatory, or concerning in any way, then follow-up must occur. Follow-up could include consultations with human resources advisors, legal counsel, and with the candidate themselves. It is permissible and legal to ask them about public postings that they have made.

Finally, posts by candidates to open-source repositories or developer exchanges should be checked. These can often indicate much about the candidate's technical competency and communications skills.

8.2.8   AGILE DEVELOPMENT TEAMS

8.2.8.1   How Do I Manage Agile Development Teams?

To answer this question, we need to discuss the Agile Manifesto. The Agile Manifesto is a document that lays the groundwork for all agile development methodologies. Its authors include many notable pioneers of object technology including Kent Beck (JUnit with Eric Gamma), Alistair Cockburn (Crystal), Ward Cunningham (Wiki, CRC cards), Martin Fowler (many books on XP, patterns, UML), Robert C. Martin (Agile, UML, patterns), and Ken Schwaber (Scrum).

The Agile Manifesto

- Our highest priority is to satisfy the customer through early and continuous delivery of valuable software.
- *Welcome changing requirements, even late in development.* Agile processes harness change for the customer's competitive advantage.
- Deliver working software frequently, from a couple of weeks to a couple of months, with a preference for the shorter timescale.
- Businesspeople and developers must work together daily throughout the project.
- Build projects around motivated individuals. Give them the environment and support they need and trust them to get the job done.
- The most efficient and effective method of conveying information to and within a development team is face-to-face conversation.

- Working software is the primary measure of progress.
- Agile processes promote sustainable development. *The sponsors, developers, and users should be able to maintain a constant pace indefinitely.*
- Continuous attention to technical excellence and good design enhances agility.
- Simplicity—the art of maximizing the amount of work not done—is essential.
- The best architectures, requirements, and designs emerge from self-organizing teams.
- *At regular intervals, the team reflects on how to become more effective,* then tunes and adjusts its behavior accordingly.

(Beck et al., 2006)

Notice the portions in italics (our emphasis), which provide specific management advice.

8.2.8.2   OK, So What Does the Agile Manifesto Have to Do with Managing Agile Teams?

The Agile Manifesto is ready-made advice for managers. It implies that managing agile teams is fun (Theory Y) and that the best outcomes arise from giving project teams what they need and leaving them alone.

8.2.8.3   Does This Approach Always Work?

No. Agile methods require much more autonomy than many managers are willing to give. More importantly, however, not everyone fits the agile methodology—a Theory X type worker will not thrive well with it. Finally, as we discussed in Chapter 2, agile methodologies do not work in every environment or with every project.

8.2.8.4   Do You Have Some More Specific Advice for Managing Agile Teams?

Don Reifer (2002b) gives some excellent advice on managing agile teams. He recommends that you clearly define what "agile methods" means upfront because, as we saw, there are many misconceptions. Then build a business case for agile methods using "hard" data. Reifer also notes that when adopting agile methods, recognize that you are changing the way your organization does business.

So, in order to be successful, you need to provide staff with support for making the transition. That support should include startup guidelines, "how-to" checklists, and measurement wizards; a knowledge base of past experience accessible by all; and education and training, including distance education and self-study courses (Reifer, 2002b).

8.3   PROJECT MANAGEMENT BASICS

8.3.1   What Is a Project?

A project is a set of tasks with a defined beginning and end. Without a defined beginning, there is no way to begin measuring progress. Without a defined end, there is no way to determine if the project has been completed, and thus progress toward completion cannot be measured. The simple project definition is recursive in that any project probably consists of more than one sub-project.

8.3.2 WHAT MAKES A SOFTWARE PROJECT DIFFERENT FROM ANY OTHER KIND OF PROJECT?

Throughout the text, various properties of software have been discussed. What has been infrequently noted, however, is that the things that make software different from other types of endeavors also make it harder to manage the software process. For example, unlike hardware to a large extent, software designers build software knowing that it will have to change. Hence, the designer has to think about both the design and redesign. That adds a level of complexity.

Of course, software development involves novelty, which introduces uncertainty. It can be argued that there is a higher degree of novelty in software than in other forms of engineering.

The uniqueness of software project management is intensified by a number of specialized activities. These include:

- The process of software development
- The complex software maintenance process
- The unique, and not-well-evolved, process of verification and validation
- The interplay of hardware and software
- The uniqueness of the software

8.3.3 IS SOFTWARE PROJECT MANAGEMENT SIMILAR TO SYSTEMS PROJECT MANAGEMENT?

Many software engineering project management activities are different from those needed for software project management. These are summarized in Table 8.6. This framework provides a model of discussion for the rest of this chapter.

8.3.4 WHAT DOES THE SOFTWARE PROJECT MANAGER CONTROL?

Software project managers may have one or more of the following elements under their control:

- Resources
- Schedule
- Functionality

Note that we say "may have" control over. It could be that the project manager controls none or only one of these. Obviously, the extent to which the manager has control indicates the relative freedom to maneuver.

Note that there is one aspect here that every manager can control—himself. That is, you have control over your reactions to situations and your attitude. These must be positive if you expect positive responses from others.

TABLE 8.6
Software Process Planning versus Project Planning

Software Engineering Planning Activities	Software Project Management Planning Activities
Determine tasks to be done	Determine skills needed for the task
Establish task precedence	Establish project schedule
Determine level of effort in person months	Determine cost of effort
Determine technical approach to solving problem	Determine managerial approach to monitoring project status
Select analysis and design tools	Select planning tools
Determine technical risks	Determine management risks
Determine process model	Determine process model
Update plans when the requirements or development environment change	Update plans when the managerial conditions and environment change

Source: Adapted from Thayer, 2002.

8.3.5 What Do You Mean by Resources?

Resources can include software, equipment, and staffing, and money to acquire more of them. There are always financial limitations, and generally, these are fixed prior to the start of the project. Many times the financing constraints change during the project.

8.3.6 What about the Schedule?

The manager should have some control over the schedule. Even if the delivery date of the product is hard, there should be some flexibility in the schedule that does not change the delivery date.

8.3.7 What about Product Functionality?

The product functionality may or may not be controllable. Often, when negotiating a project, the project manager cannot increase costs or reduce delivery time, but they can decrease product functionality in order to meet a customer's budget or schedule.

8.3.8 How Does the Project Manager Put All of These Control Factors Together?

Generally, in terms of controlling the project, the manager must understand the project goals and objectives. Next, the manager needs to understand the constraints imposed on the resources. These include cost and time limitations, performance constraints, and available staff resources. Finally, the manager develops a plan that enables them to meet the objectives within the given constraints.

Of course, monitoring and control mechanisms must be in place, including metrics. The manager should be prepared to modify the plan as it progresses. These modifications need to be made to the plan, and then team members can adjust as necessary and appropriate. Finally, a calm, productive, and positive environment is desirable to maximize the performance of the team and to keep the customer happy and confident that the job is being done right.

8.4 ANTIPATTERNS IN ORGANIZATION MANAGEMENT

	Quick Access to Antipatterns Test https://phil.laplante.io/antipatterns.php

8.4.1 WHAT DO YOU MEAN BY ANTIPATTERNS IN ORGANIZATIONS?

In troubled organizations, the main obstacle to success is frequently accurate problem identification. Diagnosing organizational dysfunction is quite important in dealing with the underlying problems that will lead to requirements engineering problems.

Conversely, when problems are correctly identified, they can almost always be dealt with appropriately. But organizational inertia frequently clouds the situation or makes it easier to do the wrong thing rather than the right thing. So how can you know what the right thing is if you've got the problem wrong?

In their groundbreaking book, Brown et al. (1998) described a taxonomy of problems or antipatterns that can occur in software architecture and design, and in the management of software projects. They also described solutions or refactorings for these situations. The benefit of providing such a taxonomy is that it assists in the rapid and correct identification of problem situations, provides a playbook for addressing the problems, and provides some relief to the beleaguered employees in these situations in that they can take consolation in the fact that they are not alone.

These antipatterns bubble up from the individual manager through organizational dysfunction and can manifest in badly stated, incomplete, incorrect, or intentionally disruptive requirements. The antipattern set consists of an almost even split of 28 environmental (organizational) and 21 management antipatterns.

8.4.2 WHAT DO YOU MEAN BY MANAGEMENT ANTIPATTERNS?

Management antipatterns are caused by an individual manager or management team ("the management"). These antipatterns address issues in supervisors that lack the talent or temperament to lead a group, department, or organization.

8.4.3 CAN YOU GIVE SOME EXAMPLES OF THE MANAGEMENT ANTIPATTERNS?

Metric Abuse and Mushroom Management are both examples of management antipatterns.

8.4.4 What Do You Mean by Metric Abuse?

The first management antipattern that might arise in requirements engineering is metric abuse, that is, the misuse of metrics either through incompetence or with deliberate malice (Dekkers & McQuaid, 2002).

At the core of many process improvement efforts is the introduction of a measurement program. In fact, sometimes the measurement program *is* the process improvement. That is to say, some people misunderstand the role measurement plays in management and misconstrue its mere presence as an improvement. This is not a correct assumption. When the data used in the metric are incorrect or the metric is measuring the wrong thing, the decisions made based upon them are likely the wrong ones and will do more harm than good.

Of course, the significant problems that can arise from metric abuse depend on the root of the problem: incompetence or malice. Incompetent metrics abuse arises from failing to understand the difference between causality and correlation; misinterpreting indirect measures; underestimating the effect of a measurement program. Here's an example of the origin of such a problem. Suppose a fire control system for a factory is required to dispense fire retardant in the event of a fire. Fire can be detected in a number of ways—based on temperature, the presence of smoke, the absence of oxygen in the presence of gases from combustion, and so on. So, which of these should be measured to determine if there is a fire? Selecting the wrong metric can lead to a case of metrics abuse.

Malicious metrics abuse derives from selecting metrics that support or decry a particular position based upon a personal agenda. For example, suppose a manager institutes a policy that tracks the number of requirements written per engineer per day and builds a compensation algorithm around this metric. Such an approach is simplistic and does not consider the varying difficulties in eliciting, analyzing, agreeing, and writing different kinds of requirements. In fact, the policy may have been created entirely to single out an individual who may be working meticulously, but too slowly for the manager's preference.

The solution or refactoring for metrics abuse is to stop the offending measurements. Measuring nothing is better than measuring the wrong thing. When data are available, people use them in decision making, regardless of their accuracy. Once the situation is reset, Dekkers and McQuaid (2002) suggest a number of steps necessary for the introduction of a meaningful measurement program:

1. Define measurement objectives and plans—perhaps by applying the goal question metric (GQM) paradigm.
2. Make measurement part of the process—don't treat it like another project that might get its budget cut or that one day you hope to complete.
3. Gain a thorough understanding of measurement—be sure you understand direct and indirect metrics, causality vs. correlation, and, most importantly, that metrics must be interpreted and acted upon.
4. Focus on cultural issues—a measurement program will affect the organization's culture; expect it and plan for it.

5. Create a safe environment to collect and report true data—remember that without a good rationale people will be suspicious of new metrics, fearful of a time-and-motion study in sheep's clothing.
6. Cultivate a predisposition to change—the metrics will reveal deficiencies and inefficiencies so be ready to make improvements.
7. Develop a complementary suite of measures—responding to an individual metric in isolation can have negative side effects. A suite of metrics lowers this risk.

If you believe that you are being metric mismanaged, then you can try to instigate the above process by questioning management about why the metrics are being collected, how they are being used, and whether there is any justification for such use. You can also offer to provide a corrective understanding of the metrics with opportunities for alternate metrics and appropriate use or more appropriate uses of the existing metrics.

8.4.5 WHAT DO YOU MEAN BY MUSHROOM MANAGEMENT?

Mushroom management is a situation in which management fails to communicate effectively with staff. Essentially, information is deliberately withheld in order to keep everyone "fat, dumb, and happy." The name is derived from the fact that mushrooms thrive in darkness and dim light but will die in the sunshine. As the old saying goes "keep them in the dark, feed them dung, watch them grow ... and then cut off their heads when you are done with them."

The dysfunction occurs when members of the team don't really understand the big picture; the effects can be significant, particularly with respect to requirements engineering when stakeholders get left out. It is somewhat insulting to assume that someone working on the front lines doesn't have a need to understand the bigger picture. Moreover, those who are working directly with customers, for example, might have excellent ideas that may have a sweeping impact on the company. So, mushroom management can lead to low employee morale, turnover, missed opportunities, and general failure.

Those eager to perpetuate mushroom management will find excuses for not revealing information, strategy, and data. To refactor this situation some simple strategies to employ include:

- Take ownership of problems that allow you to demand more transparency.
- Seek out information on your own. It's out there. You just have to work harder to find it and you may have to put together the pieces. Between you and the other mushrooms, you might be able to see most of the larger picture.
- Advocate for conversion to a culture of open-book management. With all refactoring, courage and patience are needed to effect change.

8.4.6 WHAT DO YOU MEAN BY ENVIRONMENTAL ANTIPATTERNS?

Environmental antipatterns are caused by a prevailing culture or social model. These antipatterns are the result of misguided corporate strategy or uncontrolled socio-political forces.

8.4.7   Can You Give Some Examples of Environmental Antipatterns?

Divergent Goals and *process clash* are both examples of environmental antipatterns.

8.4.8   What Do You Mean by Divergent Goals?

Everyone must pull in the same direction. There is no room for individual or hidden agendas that don't align with those of the business. The divergent goals antipattern exists when there are those who pull in different directions.

There are several direct and indirect problems with divergent goals:

- Hidden and personal agendas divergent to the mission of an organization starve resources from strategically important tasks.
- Organizations become fractured as cliques form to promote their own self-interests.
- Decisions are second-guessed and subject to "review by the replay official" as staff try to decipher genuine motives for edicts and changes.
- Strategic goals are hard enough to attain when everyone is working toward them; without complete support they become impossible and introduce risk to the organization.

There is a strong correspondence between stakeholder dissonance and divergent goals, so be very aware of the existence of both.

Since divergent goals can arise accidentally and intentionally there are two sets of solutions or refactorings.

Dealing with the first problem of comprehension and communication involves explaining the impact of day-to-day decisions on larger objectives

The second problem of intentionally charting an opposing course is far more insidious, however, and requires considerable intervention and oversight. The starting point is to recognize the disconnect between their personal goals and those of the organization. Why do they feel that the stated goals are incorrect? If the motives really are personal, that they feel their personal success cannot come with success of the organization, radical changes are needed. Otherwise the best recourse is to get them to buy into the organizational goals. This is most easily achieved if every stakeholder is represented in the definition and dissemination of the core mission and goals, and subsequently kept informed, updated, and represented.

8.4.9   What Is Meant by Process Clash?

A process clash is the friction that can arise when advocates of different processes must work together without a proven hybrid process being defined. The dysfunction appears when organizations have two or more well-intended but noncomplementary processes; a great deal of discomfort can be created for those involved. Process clash can arise when functional groups or companies (with different processes) merge, or when management decides to suddenly introduce a new process to replace an old one.

Symptoms of this antipattern include poor communications—even hostility—high turnover, and low productivity.

The solution to a process clash involves developing a hybridized approach—one that resolves the differences at the processes' interfaces. Retraining and cross-training could also be used. For example, by training all engineering groups in requirements engineering principles and practices, mutual understanding can be achieved. Another solution is to change to a third process that resolves the conflict.

8.5 TRACKING AND REPORTING PROGRESS

8.5.1 WHAT IS A WORK BREAKDOWN STRUCTURE AND WHY IS IT IMPORTANT IN PROJECT TRACKING?

The work breakdown structure (WBS) is used to decompose the functionality of the software in a hierarchical fashion. The WBS can be used for costing and project management and it forms the basis for process tracking and cost determination. The WBS consists of an outline listing of project deliverables (or phases of the project) organized hierarchically.

Figure 8.3 illustrates a simple example for the software engineering effort for the baggage inspection system. A portion of the SRS is shown.

Each organization uses its own terminology for classifying WBS components according to their level in the corporate hierarchy. The WBS may also be organized around deliverables or phases of the project life cycle. In this case, higher levels generally are performed by groups while the lowest levels are performed by individuals. A WBS that emphasizes deliverables does not necessarily specify activities.

8.5.2 WHAT IS THE LEVEL OF DETAIL OF THE TASKS IN THE WBS?

Breaking down a project into its component parts facilitates resource allocation and the assignment of individual responsibilities. But care should be taken to use a proper level of detail when creating the WBS. A very high level of detail is likely to result in

```
1.1 Software Systems Engineering
    1.1.1 Support to Systems Engineering
    1.1.2 Software Engineering Trade Studies
    1.1.3 Requirement Analysis (System)
    1.1.4 Requirement Analysis (Software)
    1.1.5 Equations Analysis
    1.1.6 Interface Analysis
    1.1.7 Support to System Test

1.2 Software Development
    1.2.1 Deliverable Software
        1.2.1.1 Requirement Analysis
        1.2.1.2 Architectural Design
```

FIGURE 8.3 High-level work breakdown structure for the baggage inspection system

micromanagement. Too low a level of detail and the tasks may become too large to manage effectively. Generally, we like to define tasks so that their duration is between several days and a few months.

8.5.3 WHAT IS THE ROLE OF THE WBS IN PROJECT PLANNING?

The work breakdown structure is the foundation of project planning. It is developed before dependencies are identified and activity durations are estimated. The WBS can also be used to identify the tasks to be used in other project management tools.

8.5.4 ARE THERE ANY DRAWBACKS TO THE TRADITIONAL WBS?

Yes, the WBS is closely associated with the waterfall model, although it can be used with other life cycle models. The WBS can have the tendency to drive the software architecture; for instance, the modular decomposition looks exactly like the WBS.

8.5.5 ARE THERE ANY ALTERNATIVES TO USING THE WBS?

No. Some project managers try to go directly to scheduling without using a WBS. In other cases, project managers try to utilize use cases as the basis of project management, but we do not recommend this approach.

8.5.6 HOW ARE WORK AND PROGRESS TRACKED IN SOFTWARE PROJECTS?

Tracking progress is important for identifying problems early, for reporting purposes, and to perform appropriate resource allocation and reallocation as required. Three tools that can help the software project manager to measure progress of a project are:

- The Gantt chart
- The critical path method (CPM)
- The program evaluation and review technique (PERT).

There are numerous commercial implementations of these tools, which typically can convert from one to the other and integrate with many popular word processing, spreadsheet, and presentation software packages.

8.5.7 WHAT IS A GANTT CHART?

Henry Gantt developed the Gantt chart during World War I for use as a planning tool. This widely used tool is simple in that it lists project tasks in a sequential and parallel fashion.

8.5.8 WHAT DOES THE GANTT CHART LOOK LIKE?

Consider the Gantt chart shown in Figure 8.4 for the baggage inspection system. Project tasks are listed along the left-hand side of the chart in a hierarchical fashion.

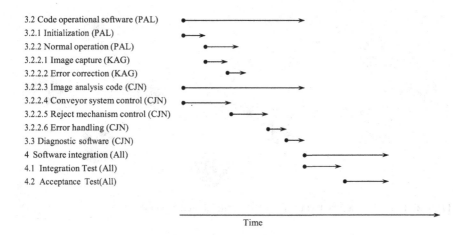

FIGURE 8.4   Partial Gantt chart for the baggage inspection system

If a work breakdown structure was used in the SDD, then it can be transferred to the chart.

A timeline is drawn along the bottom edge of the chart. Here the time units are omitted but would usually be represented by tick marks in units of days, weeks, or months. Each project subtask activity is represented by a directed arrow. The starting point of the arc is placed at the point in the timeline where the task would commence. Project durations are represented by the length of the arcs. Personnel is listed next to the project activity on the left-hand side. Milestones can be marked, and task slippage can be denoted by dashed lines in the activity arcs. The chart is updated as the project unfolds.

It can be seen from Figure 8.4 that parallel tasks can be identified and sequencing can be easily depicted. Task assignments can be made by writing the name of the responsible person next to each task. PAL, CJN, and KAG are the initials of the persons assigned to the tasks, "All" represents that all team members are involved in the task.

8.5.9   CAN THE GANTT CHART BE USED FOR LARGE PROJECTS?

For larger projects, the tasks can be broken down into subtasks, each having their own Gantt charts to maintain readability.

8.5.10   HOW CAN THE GANTT CHART BE USED FOR ONGOING PROJECT MANAGEMENT?

The strength of the Gantt chart is its capability to display the status of each activity at a glance. So long as the chart is a realistic reflection of the situation, the manager can use it to track progress, adjust the schedule, and perhaps most importantly, communicate the status of the project.

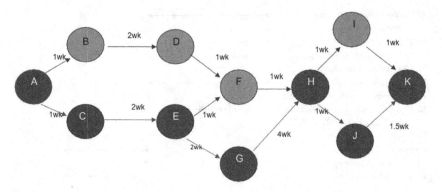

FIGURE 8.5   A generic CPM chart with the critical path highlighted

8.5.11   What Is the CPM?

The CPM (critical path method) is an improvement on the Gantt chart in that task dependencies can be more easily depicted, and task times can be represented numerically rather than visually. The method was developed in the 1950s by researchers at DuPont and Remington Rand.

The CPM chart is essentially a precedence graph connecting tasks and illustrating their dependencies along with the budgeted completion time and maximum cumulative completion time along the path from the origin to the current task (see Figure 8.5).

For example, in Figure 8.5 the tasks are A, B, C, …, K. Task A is the initial task followed by tasks B and C, which cannot start until A is completed. The time to complete task A is 1 week and 2 weeks for B and C.

8.5.12   What Are the Steps in CPM Planning?

First, specify the individual activities, which can be obtained from the work breakdown structure. This listing can be used as the basis for adding sequence and duration information in later steps.

Next, determine the sequence of those activities, including any dependencies. Note that some activities are dependent upon the completion of others. A listing of the immediate predecessors of each activity is useful for constructing the CPM network diagram.

Now draw a network diagram. CPM was originally developed as an activity on a node network, but some project planners prefer to specify the activities on the arcs.

Next, estimate the completion time for each activity. The time required to complete each activity can be estimated using past experience or the estimates of knowledgeable persons. CPM is a deterministic model that does not consider variation in the completion time, so only one number is used for an activity's time estimate.

Then identify the critical path, which is the path through the project network that has the greatest aggregate completion time.

Finally, update the CPM diagram as the project progresses because the actual task completion times will then be determined and the network diagram can be updated to include this information. A new critical path may emerge, and structural changes may be made in the network if project requirements change.

8.5.13 CAN YOU ILLUSTRATE THE TECHNIQUE USING THE BAGGAGE INSPECTION SYSTEM?

Returning to the baggage inspection system example, consider the tasks in the work breakdown structure by their numerical coding shown in the Gantt chart. These tasks are depicted in Figure 8.6.

Here tasks 3.2.1, 3.2.2.3, and 3.2.2.4 can begin simultaneously. Assume that task 3.2.1 is expected to take 4 time units (days). Notice that, for example, the arc from task 3.2.1 is labeled with "4/4" because the estimated time for that task is 4 days, and the cumulative time along that path up to that node is 4 days. Looking at task 3.2.2.1, which succeeds task 3.2.1, we see that the edge is labeled with "4/8". This is because a completion time for task 3.2.2.1 is estimated at 4 days, but the cumulative time for that path (from tasks 3.2.1 through 3.2.2.1) is estimated to be a total of 8 days.

Moving along the same path, task 3.2.2.1 is also expected to take 4 days, so the cumulative time along the path is 8 days. Finally, task 3.2.2.2 is expected to take 3 days, and hence the cumulative completion time is 11 days. On the other hand, task 3.2.2.3 is expected to take 17 days. The task durations are based on either estimation or using a tool such as constructive cost model (COCOMO), which will be discussed later. If the Gantt chart accompanies the CPM diagram, the task durations represented by the length of the arrows on the Gantt chart should correspond to those labeled on the CPM chart.

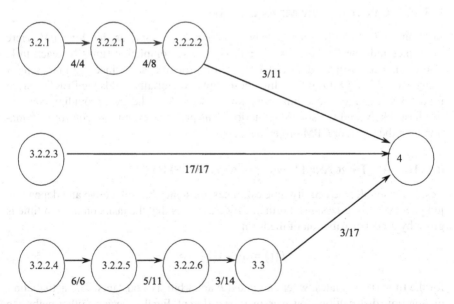

FIGURE 8.6 Partial CPM corresponding to the baggage inspection system Gantt chart shown in Figure 8.4

Moving along the last path at the bottom of Figure 8.6, it can be seen that the cumulative completion time is 17. Therefore, in this case, the two lower task paths represent critical paths. Hence, only by reducing the completion time of both lower task paths can the project completion be accelerated.

8.5.14 ARE THERE DOWNSIDES TO USING CPM?

CPM was developed for complex but fairly routine projects with minimal uncertainty in the project completion times. For less routine projects, there is more uncertainty in the completion times and this uncertainty limits the usefulness of the deterministic CPM model. An alternative to CPM is the PERT project planning model, which allows a range of durations to be specified for each activity.

8.5.15 WHAT IS PERT?

PERT (program evaluation and review technique) was developed by the U.S. Navy and Lockheed (now Lockheed Martin) in the 1950s, around the same time as CPM. PERT is identical to CPM topologically, except that PERT depicts optimistic, likely, and pessimistic completion times along each arc.

8.5.16 HOW DO YOU BUILD THE PERT DIAGRAM?

The steps are the same as for CPM except that, when you determine the estimated time for each activity, optimistic, likely, and pessimistic times are determined.

8.5.17 CAN YOU SHOW ME AN EXAMPLE?

In Figure 8.7, it can be seen that the topology is the same as that for CPM. Here the triples indicate the best, likely, and worst-case completion times for each task. These times are estimated, as in CPM, either through best engineering judgment or using a tool like COCOMO. Adding these triples vectorially yields the PERT chart in Figure 8.8. The aggregated times can now be seen along the arcs, providing cumulative best, likely, and worst-case scenarios. This provides even more control information than the Gantt or CPM project representations.

8.5.18 ARE THERE ANY DOWNSIDES TO USING PERT?

Yes. For example, the activity time estimates are somewhat subjective and depend on judgment (that is, guessing). Further, PERT assumes that the task completion time is given by a beta distribution of the form

$$f(t) = kta - 1(1 - t)b - 1$$

for the time (t) estimates, where k, a, and b are arbitrary constants. The actual completion time distribution, however, may be different. Finally, because other paths can become the critical path if their associated activities are delayed, PERT consistently underestimates the expected project completion time.

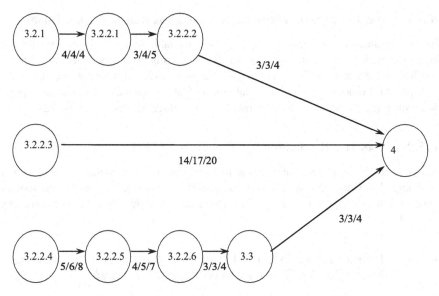

FIGURE 8.7 Partial PERT chart for the baggage inspection system showing best/likely/worst-case completion times for each task

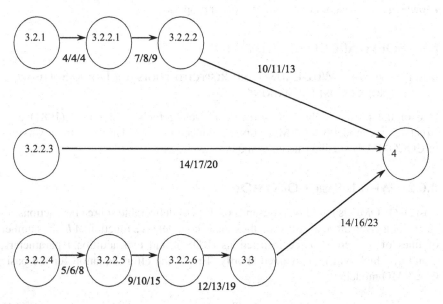

FIGURE 8.8 Partial PERT chart for the baggage handling system showing cumulative best/likely/worst-case completion times for each task

8.5.19　Are Commercial Products Available for Building These Charts?

Several commercial and open-source tools are available to develop work breakdown structures and create Gantt, CPM, and PERT diagrams. Some tools also provide to-do lists, linking tasks and dependencies, role- or skill-based tracking, and resource-leveling. Still other tools provide collaboration features such as project status reports accessible via a Web page and integrated email or threaded discussion boards.

8.5.20　Can You Recommend the Best Tool to Use?

It is not our place to recommend one implementation over another—any one will probably do. Moreover, there is no "managerial advantage" in using one tool or another. The "managerial advantage" is in the skill and knowledge of the person using the tool.

8.5.21　Is Becoming an Expert in Using the Project Planning Tools the Key to Being a Good Software Project Manager?

Absolutely not. Frankly, it is naïve to imply that mastery of a tool is a foundation for excellence in some professions. Does mastery of a word processor make you a great writer? Does learning how to use a spreadsheet program make you a financial whiz? Of course not. You need the tools to do your job, of course, but you need to have the knowledge and wisdom to use those tools appropriately.

8.6　SOFTWARE COST ESTIMATION

8.6.1　Are There Well-Known and Respected Tools for Doing Software Project Cost Estimation?

One of the most widely used software modeling tools is Boehm's (1981) constructive cost model (COCOMO), first introduced in 1981. There are three versions of COCOMO: basic, intermediate, and detailed.

8.6.2　What Is Basic COCOMO?

Basic COCOMO is based on thousands of lines of deliverable source instructions. In short, for a given piece of software, the time to complete is a function of L, the number of lines of delivered source instructions (KDSI), and two additional parameters, a and b, which will be explained shortly. This is the effort equation for the basic COCOMO model:

$$T = aL^b \tag{8.1}$$

Dividing T by a known productivity factor, in KDSI per person month, yields the number of person months estimated to complete the project. The parameters a and b are a function of the type of software system to be constructed.

For example, if the system is organic (i.e., one that is not deeply embedded in specialized hardware), then the following parameters are used: $a = 3.2$, $b = 1.05$. If the system is semidetached (i.e., partially embedded), then the following parameters are used: $a = 3.0$, $b = 1.12$.

Finally, if the system is embedded like the baggage inspection system, then the following parameters are used: $a = 2.8$, $b = 1.20$. Note that the exponent for the embedded system is the highest, leading to the longest time to complete for an equivalent number of DSI.

8.6.3   CAN YOU GIVE ME AN EXAMPLE USING COCOMO?

Suppose it is estimated somehow that the baggage inspection system will require 40 KDSI of new C code to complete. Hence, an effort level estimate of:

$$T = 2.8 \times (40K)^{1.2} = 234K$$

is obtained using COCOMO.

Suppose, then, it is known that an efficient programmer can generate 2000 LOC per month. Then, superficially at least, it might be estimated that the project would take approximately 117 person months to complete. Not counting dependencies in the task graph, implies that a five-person team would take approximately 20 months to complete the project. It would be expected, however, that more time would be needed because of task dependencies (identified, for example, using PERT).

8.6.4   WHERE DO THE SOURCE CODE ESTIMATES COME FROM?

These can come from function, feature, or use case point calculations; be based on an analysis of similar types of recently completed projects; or be provided by expert opinions.

8.6.5   SHOULD I USE MORE THAN ONE ESTIMATE?

Yes. You should use at least two methods to estimate KDSI. You can take a weighted average of the two with the weights based on your certainty of the estimate.

8.6.6   WHAT ABOUT THE INTERMEDIATE AND DETAILED COCOMO MODELS?

The intermediate or detailed COCOMO models dictate the kinds of adjustments used. Consider the intermediate model, for example. Once the effort level for the basic model is computed based on the appropriate parameters and number of source instructions, other adjustments can be made based on additional factors.

In this case, for example, if the source code estimate consists of design-modified code, code-modified, and integration-modified rather than straight code, a linear combination of these relative percentages is used to create an adaptation adjustment factor as follows.

Adjustments are then made to T based on two sets of factors, the adaptation adjustment factor, A, and the effort adjustment factor, E.

8.6.7   WHAT IS THE ADAPTATION ADJUSTMENT FACTOR?

The adaptation adjustment factor is a measure of the kind and proportion of code that is to be used in the system, namely, design-modified, code-modified, and integration-modified. The adaptation factor, A, is given by Equation 8.2.

$$A = 0.4(\%\text{design-modified}) + 0.03(\%\text{code-modified}) \\ + 0.3(\%\text{integration-modified}) \tag{8.2}$$

For new components $A = 100$. On the other hand, if all of the code is design-modified, then $A = 40$ and so on. Then the new estimation for delivered source instructions, E, is given as

$$E = L \times A / 100 \tag{8.3}$$

8.6.8   WHAT IS THE EFFORT ADJUSTMENT FACTOR?

An additional adjustment, the effort adjustment factor, can be made to the number of DSI based on a variety of factors including:

- Product attributes
- Computer attributes
- Personnel attributes
- Project attributes.

Each of these attributes is assigned a number based on an assessment that rates them on a relative scale. Then, a simple linear combination of the attribute numbers is formed based on project type. This gives a new adjustment factor, E'.

The second adjustment, effort adjustment factor, E'', is then made based on the formula

$$E'' = E' \times E \tag{8.4}$$

Then the DSI are adjusted, yielding the new effort equation

$$T = aE''^{b} \tag{8.5}$$

The detailed model differs from the intermediate model in that different effort multipliers are used for each phase of the software life cycle.

8.6.9   WHAT DO THESE ADJUSTMENT FACTORS LOOK LIKE?

Table 8.7 lists the adjustment factors corresponding to various product attributes.

TABLE 8.7
Attribute Adjustment Factors for Intermediate COCOMO

	Very Low	Low	Nominal	High		Very High	Extra High
Product attributes							
Required software reliability	0.75	0.88	1.00	1.15		1.40	
Size of application database		0.94	1.00	1.08		1.16	
Complexity of the product	0.70	0.85	1.00	1.15		1.30	1.65
Hardware attributes							
Runtime performance constraints			1.00	1.11		1.30	1.66
Memory constraints			1.00	1.06		1.21	1.56
Volatility of the virtual machine environment		0.87	1.00	1.15		1.30	
Required turnabout time		0.87	1.00	1.07		1.15	
Personnel attributes							
Analyst capability	1.46	1.19	1.00	0.86	0.71		
Software engineer capability	1.29	1.13	1.00	0.91	0.82		
Applications experience	1.42	1.17	1.00	0.86	0.70		
Virtual machine experience	1.21	1.10	1.00	0.90			
Programming language experience	1.14	1.07	1.00	0.95			
Project attributes							
Use of software tools	1.24	1.10	1.00	0.91	0.82		
Application of software engineering methods	1.24	1.10	1.00	0.91	0.83		
Required development schedule	1.23	1.08	1.00	1.04	1.10		

Note: Entries are empty if they are not applicable to the model.
Source: Adapted from Wikipedia.org.

8.6.10   How Widely Used Is COCOMO?

COCOMO is widely recognized and respected as a software project management tool. It is useful even if the underlying model is not really understood. COCOMO software is commercially available and can be found on the Web for free.

8.6.11   What Are the Downsides to Using COCOMO?

One drawback is that the model does not consider the leveraging effect of productivity tools. The model also bases its estimation almost entirely on LOC, not on program attributes, which is something that FPs do. FPs, however, can be converted to source code estimates using standard conversion formulas.

8.6.12   What Is COCOMO II?

COCOMO II is a major revision of COCOMO that is evolving to deal with some of the previously described shortcomings. For example, the original COCOMO 81 model was defined in terms of DSI. COCOMO II uses the metric SLOC instead of DSI. The new model helps better accommodate more expressive modern languages as well as software generation tools that tend to produce more code with essentially the same effort.

In addition, some of the more important factors that contribute to a project's expected duration and cost are included in COCOMO II as new scale drivers. These five scale drivers are used to modify the exponent in the effort equation:

- Precedentedness (novelty of the project)
- Development flexibility
- Architectural/risk resolution
- Team cohesion
- Process maturity.

 Quick Access to COCOMO II – Constructive Cost Model calculator https://phil.laplante.io/requirements/COCOMO.php

The first two drivers, precedentedness and development flexibility, describe many of the same influences found in the adjustment factors of COCOMO 81.

8.6.13   What Is WEBMO?

WEBMO is a derivative of COCOMO II that is geared specifically to project estimation of web-based projects (it has been reported that COCOMO is not a good predictor in some cases). WEBMO is based on a different set of predictors, namely,

- Number of function points
- Number of xml, html, and query language links
- Number of multimedia files
- Number of scripts
- Number of Web building blocks.

For WEBMO, the effort and duration equations are:

$$\text{Effort} = A\prod_{i=1}^{9} cd_i \text{size}^{P_1}$$

$$\text{Duration} = B(\text{Effort})^{P_2}$$

(8.7)

where A and B are constants, P_1 and P_2 depend on the application domain, and cd_i are cost drivers based on:

- Product reliability and complexity
- Platform difficulty
- Personal capabilities
- Personal experience
- Facilities
- Schedule constraints
- Degree of planned reuse
- Teamwork
- Process efficiency.

with qualitative ratings ranging from very low to very high and numerical equivalents shown in Table 8.8 (Reifer, 2002a).

TABLE 8.8
WEBMO Cost Drivers and Their Values

	Ratings				
	Very Low	Low	Nominal	High	Very High
Cost driver					
Product reliability	0.63	0.85	1.0	1.30	1.67
Platform difficulty	0.75	0.87	1.00	1.21	1.41
Personnel capabilities	1.55	1.35	1.00	0.75	0.58
Personnel experience	1.35	1.19	1.00	0.87	0.71
Facilities	1.35	1.13	1.00	0.85	0.68
Schedule constraints	1.35	1.15	1.00	1.05	1.10
Degree of planned reuse	—	—	1.00	1.25	1.48
Teamwork	1.45	1.31	1.00	0.75	0.62
Process efficiency	1.35	1.20	1.00	0.85	0.65

Source: Reifer, 2002a.

8.7   PROJECT COST JUSTIFICATION

8.7.1   Is Software an Investment or an Expense?

It depends on whom you ask. Many software project managers see the investment in new tools or upgrade of old ones as an investment. But the Chief Financial Officer might see the purchase as a pure expense. There is an accounting answer to this question, too, but we don't want to get into the technical details of how accountants determine whether a purchase is an expense item or a capital acquisition. The point is that many software project managers are being asked to justify their activities and purchases of software and equipment. Therefore, it is in the project manager's best interest to know how to make a business case for the activity.

8.7.2   What Is Software Return on Investment and How Is It Defined?

Return on investment (ROI) is a rather overloaded term that means different things to different people. To some it means the value of the software activity at the time it is undertaken. To some, it is the value of the activity at a later date. To some, it is just a catchword for the difference between the cost of software and the saving created from the utility of that software. Finally, to some, there is a more complex meaning.

8.7.3   What Is an Example of a Project ROI Justification?

Consider the following situation. A project manager has the option of either purchasing a new testing tool for $250,000 or using the same resources to hire and train additional testers. Currently, $1 million is budgeted for software testing. It has been projected that the new testing tool would provide $500,000 in immediate cost savings by automating several aspects of the testing effort. The effort savings would allow fewer testers to be assigned to the project. Should the manager decide to hire new testers, they would have to be hired and trained (these costs are included in the $250,000 outlay) before they can contribute to the project.[2] At the end of two years, it is expected that the new testers will be responsible for $750,000 in rework cost savings by finding defects prior to release that would not otherwise be found. The value justification question is "should the project be undertaken or not?" We can answer this question after discussing the net present value.

8.7.4   Yes, but How Do You Measure ROI?

One traditional measure of ROI for any activity, whether software related or not, is given as

$$ROI = Average\ Net\ Benefits/Initial\ Costs \tag{8.8}$$

The problem with this model for ROI is the accurate representation of average net benefits and initial costs.

8.7.5 OK, So How Can You Represent the Net Benefit and Initial Cost?

Commonly used models for the valuation of some activity or investment include net present value (NPV), internal rate of return (IRR), profitability index (PI), and payback. We will look at each of these shortly.

Other methods include Six Sigma and proprietary balanced scorecard models. These kinds of approaches seek to recognize that financial measures are not necessarily the most important component of performance. Further considerations for valuing software solutions might include customer satisfaction, employee satisfaction, and so on, which are not usually modeled with traditional financial valuation instruments.

There are other, more complex, accounting-oriented methods for valuing software. Discussion of these techniques is beyond the scope of this text. The references at the end of the chapter can be consulted for additional information; see, for example, Raffo, Settle, & Harrison (1999) and Morgan (2005).

8.7.6 What Is NPV and How Can I Use It?

NPV is a commonly used approach to determine the cost of software projects or activities. Here is how to compute NPV. Suppose that FV is some future anticipated payoff either in cash or anticipated savings. Suppose r is the discount rate[3] and Y is the number of years that the cash or savings is expected to be realized. Then the NPV of that payoff is

$$NPV = FV/(1 + r)^Y \tag{8.9}$$

NPV is an indirect measure because you are required to specify the market opportunity cost (discount rate) of the capital involved.

8.7.7 Can You Give an Example of an NPV Calculation for a Software Situation?

To see how you can use this notion as a project manager, suppose that you expect a programming staff training initiative to cost your company $60,000. You believe that benefits of this improvement initiative are expected to total $100,000 of reduced code rework two years in the future. If the discount rate is 3%, should the initiative be undertaken?

To answer this question, we calculate the NPV of the strategy, considering its cost as

$$NPV = 100,00-/1.03^2 - 60,000 = 34,259 \tag{8.10}$$

Because the NPV is positive, the project should be undertaken.

For a sequence of cash flows, CF_n, where $n = 0, ..., k$ represents the number of years from the initial investment, the NPV of that sequence is

$$NPV = \sum_{n=0}^{k...CF_n} (1+r)^n \tag{8.11}$$

CF_n could represent, for example, a sequence of related expenditures over a period of time, such as the ongoing maintenance costs or support fees for some software package.

8.7.8 WHAT IS THE ANSWER TO THE QUESTION OF ACQUIRING THE TESTING TOOL?

To figure out if we need to undertake this project, we assume an annual discount rate of 10% for ease of calculation. Now we calculate the NPV of both alternatives. The testing tool is worth $500,000 today, so its NPV is

$$PV_{tool} = \$500,000/(1.10)^0 = \$500,000 \qquad (8.12)$$

To hire testers is worth $750,000 in two years, so its NPV is

$$PV_{hire} = \$750,000/(1.10)^2 = \$619,835 \qquad (8.13)$$

Therefore, under these assumptions, the personnel hire option would be the preferred course of action. However, as the projected return goes farther into the future, it becomes more difficult to forecast the amount of the return. All sorts of things could happen—the project could be canceled, new technology could be discovered, and the original estimate of the rework could change. Thus, the risk of the project may differ accordingly.

8.7.9 WHAT IS AN IRR?

IRR is defined as the discount rate in the NPV equation that causes the calculated NPV to be zero. NPV is not the ROI. But the IRR is useful for computing the "return" because it does not require knowledge of the cost of capital.

To decide if we should undertake an initiative, we compare the computed IRR to the return of another investment alternative. If the IRR is very low, then we might simply want to take this money and find an equivalent investment with lower risk (e.g., to undertake a different corporate initiative or even simply buy bonds). But if the IRR is sufficiently high, then the decision might be worth whatever risk is involved.

8.7.10 CAN YOU GIVE AN EXAMPLE OF AN ROI CALCULATION?

Suppose the programming staff training initiative previously discussed is expected to cost $50,000. The returns of this improvement are expected to be $100,000 of reduced rework two years in the future. We would like to know the IRR on this activity.

Here, NPV = $100,000 - (1 + r)^2 - 50,000$. We now wish to find the r that makes the NPV = 0; that is, the "break-even" value. Using our IRR equation,

$$r = [100,000/50,000)]^{\frac{1}{2}} - 1 \qquad (8.14)$$

This means $r = 0.414 = 41.4\%$. This rate of return is very high, and we would likely choose to undertake this programming staff training initiative.

8.7.11 What Is a PI?

The PI is the NPV divided by the cost of the investment, I:

$$PI = NPV / I \qquad (8.15)$$

PI is a "bang-for-the-buck" measure, and it is appealing to managers who must decide between many competing investments with positive NPV financial constraints. The idea is to take the investment options with the highest PI first until the investment budget runs out. This approach is not bad but can sub-optimize the investment portfolio.

8.7.12 How Can PI Sub-optimize the Decision?

Consider the set of software investment decisions shown in Table 8.9. Suppose the capital budget is $500,000. The PI ranking technique will pick A and B first, leaving inadequate resources for C. Therefore, D will be chosen leaving the overall NPV at $610,000. However, using an integer programming approach will recommend taking projects A and C for a total NPV of $660,000.

8.7.13 Should I Use PI at All?

Yes, PI is useful in conjunction with NPV to help optimize the allocation of investment dollars across a portfolio of projects.

8.7.14 What Is Payback?

To the project manager, payback is the time it takes to get the initial investment back out of the project. Projects with short paybacks are preferred, although the term "short" is completely arbitrary. The intuitive appeal is reasonably clear: the payback period is easy to calculate, communicate, and understand.

TABLE 8.9
A Portfolio of Software Project Investment Decisions

Project	Investment (in hundreds of thousands of dollars)	NPV (in hundreds of thousands of dollars)	PI
A	200	260	1.3
B	100	130	1.3
C	300	360	1.20
D	200	220	1.1

8.7.15   How Can Payback Be Applied in a Software Project Setting?

Suppose changing vendors for a particular application software package is expected to have a switching cost of $100,000 and result in a maintenance cost savings of $50,000 per year. Then the payback period for the decision to switch vendors would be two years.

8.7.16   This Seems Simplistic. Is Payback Really Used?

Yes. Because of its simplicity, payback is the least likely ROI calculation to confuse managers. However, if the payback period is the only criterion used, then there is no recognition of any cash flow, small or large, to arrive after the cutoff period. Furthermore, there is no recognition of the opportunity cost of tying up funds. Because discussions of payback tend to coincide with discussions of risk, a short payback period usually means a lower risk. However, all criteria used in the determination of payback are arbitrary. From an accounting and practical standpoint, the discounted payback is the metric that is preferred.

8.7.17   What Is Discounted Payback?

The discounted payback is the payback period determined on discounted cash flows rather than undiscounted cash flows. This method considers the time (and risk) value of money invested. Effectively, it answers the questions "How long does it take to recover the investment?" and "what is the minimum required return?"

If the discounted payback period is finite in length, it means that the investment plus its capital costs are recovered eventually, which means that the NPV is at least as great as zero. Consequently, a criterion that says to go ahead with the project if it has *any* finite discounted payback period is consistent with the NPV rule.

8.7.18   Can You Give an Example?

In the previous PI example, there is a switching cost of $100,000 and an annual maintenance savings of $50,000. Assuming a discount rate of 3%, the discounted payback period would be longer than two years because the savings in year two would have an NPV of less than $50,000 (figure out the exact payback period for fun). But because we know that there is a finite discounted payback period, we know that we should go ahead with the initiative.

8.8   RISK MANAGEMENT

8.8.1   What Are Software Risks?

Software risks are "anything that can lead to results that deviate negatively from the stakeholders' real requirements for a project" (Gilb, 2002).

8.8.2   How Do Risks Manifest in Software?

There are two kinds of software risks: external and internal. Internal risks include requirements changes, unrealistic requirements, incorrect requirements; shortfalls in externally furnished components; problems with legacy code; and lack of appropriate resources. These kinds of risks appear to be controlled most likely by the software project manager or their organization. External risks are related to the business environment and include changes in the situation of customers, competitors, or suppliers; economic situations that change the cost structure; governmental regulations; weather, terrorism, and so on. Of course, the project manager can control none of these risks. Instead, they need to plan for them so they can be mitigated when they arise.

8.8.3   How Does the Project Manager Identify, Mitigate, and Manage Risks?

Many of the risks can be managed through close attention to the requirements specification and design processes. Prototyping (especially throwaway) is also an important tool in mitigating risk. Judicious and vigorous testing can reduce or eliminate many of these risks.

8.8.4   What Are Some Other Ways That the Software Project Manager Can Mitigate Risk?

Table 8.10, which is a variation of a set of recommendations from Boehm (1989), summarizes the risk factors and possible approaches to risk management and mitigation.

8.8.5   Is There a Predictive Model for the Likelihood of Any of These Risks?

Yes. Once again, Boehm (1989) offers us some advice on the likelihood of various kinds of risks driving up cost, shown in Table 8.11.

8.8.6   Do You Have Any Other Advice about Managing Risk in Software Projects?

Tom Gilb, a software risk management specialist, suggests that the project manager ask the following questions throughout the life of the software project (Gilb, 2002):

- Why isn't the improvement quantified?
- What is the degree of risk or uncertainty and why?
- Are you sure? If not, why not?
- Where did you get that from? How can I check it out?
- How does your idea measurably affect my goals and budgets? Did we forget anything critical to survival?

TABLE 8.10
Various Project Risk Sources and Possible Management, Measurement, Elimination, and Mitigation Techniques

Risk Factor	Possible Management/Mitigation Approach
Incomplete and imprecise specifications	Prototyping requirements reviews formal methods
Difficulties in modeling highly complex systems	Prototyping testing
Uncertainties in allocating functionality to software or hardware and subsequent turf battles	Prototyping requirements reviews
Uncertainties in cost and resource estimation	Project management metrics
Difficulties with progress monitoring	Project management monitoring tools metrics
Rapid changes in software technology and underlying hardware technology	Prototyping testing
Measuring and predicting reliability of the software	Metrics testing
Problems with interface definition	Prototyping
Problems encountered during software-software or hardware-software integration	Prototyping testing
Unrealistic schedules and budgets	Project management monitoring tools metrics
Gold plating	Code audits and walkthroughs
Shortfalls in externally furnished components	Testing
Real-time performance shortfalls	Prototyping testing
Trying to strain the limits of computer science capabilities	Code audits and walkthroughs testing

Source: Adapted from Boehm, 1989.

- How do you know it works that way? Did it before?
- Do we have a complete solution? Are all requirements satisfied?
- Are we planning to do the "profitable things" first? Who is responsible for failure or success?
- How can we be sure the plan is working during the project or earlier? Is it "no cure, no pay" in a contract? Why not?

He offers other advice that reflects the healthy skepticism that the project manager needs to have:

- Re-think the deadline given—is it for real?
- Re-think the solution—is it incompatible with the deadline?
- What is the requestor's real need/point of view?

TABLE 8.11
Probabilistic Assessment of Risk

Cost Driver	Improbable (0.0–0.3)	Probable (0.4–0.6)	Frequent (0.7–1.0)
Application	Nonreal-time, little system interdependency	Embedded, some system interdependencies	Real-time embedded, strong interdependency
Availability	In place, meets need dates	Some compatibility with need dates	Nonexistent, does not meet need dates
Configuration management	Fully controlled	Some controls	No controls
Experience	High experience ratio	Average experience ratio	Low experience ratio
Facilities	Little or no modification	Some modifications, existent	Major modifications, nonexistent
Management environment	Strong personnel management approach	Good personnel management approach	Weak personnel management approach
Mix	Good mix of software disciplines	Some disciplines inappropriately represented	Some disciplines not represented
Requirements stability	Little or no change to established baseline	Some change in baseline expected	Rapidly changing or no baseline
Resource constraints	Little or no hardware-imposed constraints	Some hardware-imposed constraints	Significant hardware-imposed constraints
Rights	Compatible with maintenance and development plans	Partial compatibility with maintenance and development plans	Incompatible with maintenance and development plans
Size	Small, noncomplex, or easily decomposed	Medium to moderate complexity, decomposable	Significant hardware-imposed constraints
Technology	Mature, existent, in-house experience	Existent, some in-house experience	New or new application, little experience

Source: Boehm, 1989.

- Don't blindly accept "expert" opinions.
- Determine which components really must be delivered at the deadline.

8.8.7 How Does Prototyping Mitigate Risk?

Prototyping gives users a feel for how well the design approach works and increases communication between those who write requirements and the developers throughout the requirements specification and design process. Prototyping can be used to exercise novel hardware that may accompany an embedded system. Prototyping can also detect problems and identify deficiencies early in the life cycle, where changes are more easily and inexpensively made.

8.8.8 Are There Risks to Software Prototyping?

Indeed, there are. For example, a prototype may not provide good information about timing characteristics and real-time performance, which lulls the designers into a false sense of security. Often the pressures of bringing a product to market lead to a temptation to carry over portions of the prototype into the final system. Therefore, use throwaway prototypes as much as possible.

8.8.9 Are There Other Ways to Discover Risks So That They Can Be Mitigated?

Yes. The best way is to ask experts who have worked on similar projects. There really is no substitute for experience.

NOTES

1　The term manager as a general term for anyone who is the responsible charge for one or more other persons developing, managing, installing, supporting, or maintaining systems and software. Other typical titles include "Software Project Manager," "Technical Lead," and "Senior Developer."
2　Such a cost is called a "sunken cost" because the money is gone whether one decides to proceed with the project or not.
3　The interest rate charged by the U.S. Federal Reserve. The cost of borrowing any capital will be higher than this base rate.

FURTHER READING

Andriole, S. (1998). The politics of requirements management. *IEEE Software, 15*, 82–84.
Beck, K. et al. (2006) *The Agile Manifesto* http://agilemanifesto.org/principles.html, accessed April 2022.
Berander, P., Damm, L.O., Eriksson, J., Gorschek, T., Henningsson, K., Jönsson, P., ... & Wohlin, C. (2005). Software quality attributes and trade-offs. *Blekinge Institute of Technology, 97*(98), 19.
Boehm, B.W. (1981). *Software Engineering Economics.* Prentice Hall, Englewood Cliffs, NJ.
Boehm, B.W. (1989). *Software Risk Management.* IEEE Computer Society Press, Los Alamitos, CA.
Boehm, B. & Turner, R. (2003). *Balancing Agility and Discipline: A Guide to the Perplexed.* Addison-Wesley, Boston, MA.
Bramson, R. (1988). *Coping with Difficult People.* Dell Paperbacks, New York.
Brooks, F.P. (1995). *The Mythical Man Month. 20th Anniversary Edition.* Addison-Wesley, Boston, MA.
Brown, W.J., Malveau, R.C., McCormick, H.W., & Mowbray, T.J. (1998). *AntiPatterns: Refactoring Software, Architectures, and Projects in Crisis.* Wiley, New York.
Cohen, C., Birkin, S., Garfield, M., & Webb, H. (2004) Managing conflict in software testing. *Commun. ACM, 47*(1), 76–81.

Covey, S.R., *Principle-Centered Leadership*, Simon & Schuster, New York, 1991.

Dekkers, C.A., & McQuaid, P.A. (2002). The dangers of using software metrics to (mis) manage. *IT Professional, 4*(2): 24–30.

Gilb, T. (2002). Risk Management: A practical toolkit for identifying. *Analyzing and coping with project risks.* www. result-planning. com.

IEEE (1998). *IEEE 1490-1998 IEEE Guide – Adoption of PMI Standard – A Guide to the Project Management Body of Knowledge.* IEEE Standards, Piscataway, NJ.

Jones, C. (1996). *Patterns of software system failure and success.* Itp-Media.

Kassab, M. (2013, May). An integrated approach of AHP and NFRs framework. In *IEEE 7th International Conference on Research Challenges in Information Science (RCIS)* (pp. 1–8). IEEE.

Kassab, M. (2014, June). Early effort estimation for quality requirements by AHP. In *International Conference on Computational Science and Its Applications*, pp. 106–118, Springer International Publishing.

Kassab, M., Destefanis, G., DeFranco, J., & Pranav, P. (2021, May). Blockchain-Engineers Wanted: an Empirical Analysis on Required Skills, Education and Experience. In *2021 IEEE/ACM 4th International Workshop on Emerging Trends in Software Engineering for Blockchain (WETSEB)* (pp. 49–55). IEEE.

Kassab, M., & Kilicay-Ergin, N. (2015). Applying analytical hierarchy process to system quality requirements prioritization. *Innovations in Systems and Software Engineering, 11*(4), 303–312.

Kassab, M. & Laplante, P.A. (2022). The Current and Evolving Landscape of Requirements Engineering State of Practice. *IEEE Software.* DOI Bookmark: 10.1109/MS.2022. 3147692

Kassab, M., Laplante, P.A., Defranco, J., Neto, V.V.G., & Destefanis, G. (2021). Exploring the Profiles of Software Testing Jobs in the United States. *IEEE Access, 9*, 68905–68916.

Laplante, P.A. (2003). *Software Engineering for Image Processing Systems*, CRC Press, Boca Raton, FL.

Laplante, P. A.; Kassab, M. (2022). *Requirements engineering for software and systems.* 4th edition. Auerbach Publications.

Morgan, J.N. (2005). A roadmap of financial measures for IT project ROI. *IT Prof.*, Jan./ Feb., 52–57.

Neill, C.J., Laplante, P.A. and DeFranco, J.F. (2012). *Antipatterns: Identification, Refactoring, and Management.* Second Edition. Taylor and Francis.

Raffo, D., Settle, J., and Harrison, W. (1999). *Investigating Financial Measures for Planning of Software IV&V.* Portland State University Research Report #TR-99-05, Portland, OR.

Reifer, D. (2000). Web development: estimating quick-to-market software. *IEEE Software*, Nov./Dec., 57–64.

Reifer, D. (2002a). Estimating web development costs: there are differences. *Crosstalk, 15*(6), 13–17.

Reifer, D. (2002b). How good are agile methods? *Software*, July/Aug., 16–18.

Stelzer, D. and Mellis, W. (1998). Success factors of organizational change in software process improvement. *Software Process—Improvement and Practice, 4*, 227–250.

Saaty, T.L. (1988). What is the analytic hierarchy process? In *Mathematical models for decision support* (pp. 109–121). Springer, Berlin, Heidelberg.

Thayer, R.H., Software system engineering: a tutorial, *Computer, 35*(4)68–73, 2002.

Wang, C., Cui, P., Daneva, M., & Kassab, M. (2018, October). Understanding what industry wants from requirements engineers: An exploration of RE jobs in Canada. In *Proceedings of the 12th ACM/IEEE International Symposium on Empirical Software Engineering and Measurement* (pp. 1–10).

9 Software Engineering
Roadmap to the Future[1]

OUTLINE

- Global software engineering
- Software engineering and small businesses
- Software engineering and disruptive technologies

9.1 INTRODUCTION

The Greek philosopher, Heraclites, is famous for his conviction that constant change is a fundamental truth of the universe. With fingers firmly on the pulse of changes to disruptive technologies and development approaches, one can envision the factors that will influence the landscape of software engineering in near future, say, ten years from now. There is no question that technological disruption (e.g., AI, IoT, Blockchain) will affect the way that businesses, consumers, or industries function. These same technologies will also influence software engineering practices.

Moreover, as systems become more and more distributed and divided into a set of components, so do the software artifacts (e.g., requirements, architecture, test cases). Globalization and outsourcing drive the implementation of market-specific features. Stakeholders can be geographically located in multiple countries and components sourced from other countries.

In addition, systems are no longer always built from scratch. Companies are transitioning to offering products as services instead of standalone offerings. The use of commercial off-the-shelf software and the move to microservices architectures and cross-organizational systems will also influence the traditional requirements engineering practices. For example, requirements will keep on being layered on top of one another due to the separation of program features into separate services; these will need to have their own set of requirements, as well as proper maintenance. Besides, requirements management will need to include an ongoing negotiation process with a continuously expanding set of stakeholders and dynamic architectures that needs to be adaptable and expandable.

Finally, the ongoing open-source movement will continue to influence software engineering as well. As engineers keep on using modules from a myriad of other developers, more focus must be paid to ensuring license compliance, and the potential of constant updates breaking the system and the packages' average lifetime.

This chapter will discuss some of the above factors that will influence the near future of software engineering.

DOI: 10.1201/9781003218647-10

9.2   GLOBAL SOFTWARE ENGINEERING

Software engineering requires collaboration-intensive activities. Global software development is increasingly becoming a popular software engineering paradigm as many companies are adopting global software development to reduce development costs. But global and even onshore outsourcing present all kinds of challenges to the software engineering endeavor. These include time delays and time zone issues, the costs and stresses of physical travel to client and vendor sites when needed, and the disadvantages of virtual conferencing and telephone. Even simple email communications cannot be relied upon entirely, even though for many globally distributed projects, informal emails and email-distributed documents are the main form of collaboration. But email use leads to frequent context switching, information fragmentation, and the loss of nonverbal cues.

When the offshoring takes place in a country with a different native language and substantially different culture, new problems may arise in terms of work schedules, work attitudes, communication barriers, and customer–vendor expectations of how to conduct business. Moreover, offshoring introduces a new risk factor: geopolitical risk—and this risk must be understood, quantified, and somehow factored into the software engineering process and schedule. There are also vast differences in laws, legal processes, and even the expectations of honesty in business transactions around the world. These issues are particularly relevant during the various software engineering phases.

9.2.1   WHAT ARE SOME OF THE PROBLEMS WHEN DEALING WITH GLOBAL SOFTWARE ENGINEERING?

Many existing studies discuss these problems (e.g., Satzger et al., 2014;). Bhat, Mayank and Santhosh (2006) highlighted nine specific problems they observed or experienced when dealing with global software engineering. These included:

- Conflicting client-vendor goals
- Low client involvement
- Conflicting requirements engineering approaches (between client and vendor)
- Misalignment of client commitment with project goals
- Disagreements in tool selection
- Communication issues
- Disowning responsibility
- Sign-off issues
- Tools misaligned with expectation

Bhat et al. (2006) suggest that the following success factors were missing in these cases, based on an analysis of their project experiences:

- Shared goal—that is, a project metaphor
- Shared culture—in the project sense, not in the sociological sense
- Shared process

- Shared responsibility
- Trust

These suggestions are largely consistent with agile methodologies, though we have already discussed the challenges and advantages of using agile approaches to requirements engineering.

9.2.2 What Is the Difference between Internationalization and Distribution?

When discussing the problem of global software engineering, for example, Schmid (2014) suggests distinguishing between internationalization and distribution. While the former refers to the development for a set of international customers (perhaps with a single, localized development team), the latter refers to the development in a globally distributed environment, where many stakeholders are in a different location from the customer(s). While both issues often co-occur, they are different and may require slightly different strategies for handling them.

9.2.3 What Are the Issues to Consider When Dealing with Internationalization?

Hofstede (2001) studied a large data set pertaining to IBM employee values that had been collected across worldwide sites between 1967 and 1973. Hofstede found that these values varied significantly depending on the site at which the employees worked. Hofstede concluded that there were four dimensions of social norms along which cultural differences between countries could be perceived: the power distance index, individualism, masculinity, and uncertainty avoidance. He later added a fifth dimension: long-term orientation.

More recently, Schmid (2014) identified eight context issues that impact the requirements of a system due to the place where the software is used (internationalization):

1. Language: Customers may use a different language than the development team.
2. User interface: Due to language issues and cultural issues, it might be necessary to create different kinds of user interfaces.
3. Local standards: The customer might use different standards like calendars or measurement systems than the developer.
4. Laws and regulations: These might be fundamentally different between customer and developer, but taken as obvious, as the customer is never concerned with other rules, thus the presence of a difference might even go easily unnoticed.
5. Cultural differences: Due to cultural differences, the expectations regarding system behavior may be profoundly different.
6. Regional issues: The situation for the customer might again be regionally subdivided, so that not a single solution, but a product line or customizable system is required.

7. Educational and work context related issues: In different regions, different levels of educational background might be expected, which may require very different user interfaces (e.g., when operating certain machinery).
8. Environmental conditions: The products might need to work in environmental conditions, fundamentally different from those the developers assume, giving rise to corresponding requirements.

9.2.4 What Are the Issues to Consider When Dealing with a Distributed Team?

The COVID-19 pandemic quickly forced companies to adjust to non-co-located work paradigms. For those companies not having such experience the distributed nature of a development team introduced many additional challenges with respect to all aspects of software development. For example, with respect to requirements engineering:

1. Elicitation problems may occur due to the distributed nature of the development
2. Communicating the captured requirements in a distributed software development environment
3. Cooperation across organizational borders to satisfy the requirements.

Similar kinds of problems can occur with respect to software design, development, testing, integration and management.

But companies have been dealing with distributed development issues for a long time and there are many studies of practice and experience in solving these problems. For example, Sangwan and Laplante (2006) studied the use of test-driven development in two large-scale global software development projects at a Fortune 500 corporation. They found that increased informal and formal communication, facilitated by appropriate change management and notification tools, were important. Test frameworks like the XUnit family were also used at the systems level. Finally, the use of commercial or open-source test workbenches and bug repositories to track test case dependencies and test failures is essential.

There are, of course, additional studies and we expect many more to emerge recounting successes, failures and best practices for the post-pandemic world that can be applied purposively in all distributed development settings.

9.2.5 How Can Tools Play a Role in Globally Distributed Software Engineering Processes?

Sinha and Sengupta (2006) suggest that software tools can play an important role in combating several of the above challenges, though there are not many appropriate tools for this purpose. Appropriate software tools must support:

• Informal collaboration
• Change management

- Promoting awareness (e.g., auto-emailing stakeholders when triggers occur)
- Managing knowledge—providing a framework for saving and associating unstructured project information.

There are several commercial and even open-source solutions that claim to provide these features, but we leave the product research of these to the reader.

9.3 SOFTWARE ENGINEERING PRACTICES IN SMALL BUSINESSES

9.3.1 WHAT IS A SMALL BUSINESS?

Small businesses are defined as enterprises that employ fewer than 50 persons and whose annual turnover or annual balance sheet total does not exceed EUR 10 million, according to the European Commission (2005). Hence, most startups may also be considered small businesses. In the United States, 88% of employer firms are considered to be small businesses (JPMorgan, 2018). These businesses accounted for over half of net job creation in the U.S. in 2014. According to the U.S. Small Business Administration, the overwhelming majority of software development firms are also considered to be small businesses, and this will most likely continue to be the case for the next decade. Hence, it is worth discussing the software engineering state of practices in small businesses.

9.3.2 WHAT CHARACTERIZES SOFTWARE ENGINEERING PRACTICES IN SMALL BUSINESSES?

Due to the small staff sizes, often the software engineer plays multiple roles at once. It is very common to find a software engineer who acts as the developer, architect, and/ or the QA. It is less likely to find sophisticated usages of processes in small businesses. Financial constraints can hamper the use of software engineering processes and tools. Typically, a small organization's attention to processes may come only after early successes that haven't relied heavily on it. In Kassab (2021), observations from the four RE surveys are reported on how requirements engineering are conducted in small businesses:

- A wide diversity in the employed requirements engineering practices for small businesses was reported in the 2020 Survey. We can rationalize that this diversity is the result of evolutionary adaptation, as these businesses have to adapt to their particular ecological niche. The context of these businesses is the essence of their adopted practices.
- Despite the diversity, we note that when requirements engineering practices are employed, there is a tendency to adopt the easy, flexible, and inexpensive techniques (e.g., the majority opted for open-source RE management tools).
- The adoption of inexpensive requirements engineering techniques doesn't necessarily hinder the level of satisfaction of employing them in small businesses. This is evidenced in a third observation when we compared the level of satisfaction in regard to the employed RE activities between the small

businesses sample and the overall sample. Even greater more satisfaction levels were reported in a small business sample in regard to the followed SDLC, requirements elicitation and analysis, and requirements traceability.

- A high degree of cultural cohesion that the small businesses exhibit with the majority of responses reporting an agreement that the team size was adequate for the RE challenges that face the project, and an agreement that the ability and previous software development experience of the software development team was adequate. The high level of agreement may imply a homogeneity in the professional environment of the analyzed sample which contributes to efficiency in communication and sharing an understanding of the requirements within a small team.

We surmise that these kinds of findings apply to other software engineering practices in small business environments.

9.4 SOFTWARE ENGINEERING AND DISRUPTIVE TECHNOLOGIES

It is impossible to address all the potential research areas of the diverse facets of software engineering or to comprehensively explore the disruptive impact of change in those areas, or even to hint at the potential for innovative creativity brought about by those same disruptions. Instead, this section will highlight a selection of disruptive technologies and their commensurate opportunities and challenges in software engineering.

9.4.1 SOFTWARE ENGINEERING AND THE INTERNET OF THINGS

9.4.1.1 What Is the Internet of Things?

The term "Internet of Things" (IoT) has recently become popular to emphasize the vision of a global infrastructure that connects physical objects/things, using the same Internet Protocol, allowing them to communicate and share information (Sula et al., 2013). The term "IoT" was coined by Kevin Ashton in 1999 to refer to "uniquely identifiable objects/things and their virtual representations in an Internet-like structure" (Han, 2011; Uzelac, Gligoric, & Krco, 2015).

According to the industry analysis firm Gartner, 8.4 billion "things" were connected to the Internet in 2017, excluding laptops, computers, tablets, and mobile phones. According to the GSM Association, an industry organization that represents the interests of mobile network operators worldwide, the number of IoT devices is expected to grow to 25.1 billion by 2025.

While there is still no universally accepted and actionable definition, Jeff Voas, a computer scientist at the U.S. National Institute of Standards and Technology (NIST), recommends using the acronym NoT (Network of Things) interchangeably to refer to IoT systems:

> IoT is an instantiation of a NoT, more specifically, IoT has its 'things' tethered to the Internet. A different type of NoT could be a Local Area Network (LAN),

with none of its 'things' connected to the Internet. Social media networks, sensor networks, and the Industrial Internet are all variants of NoTs. This differentiation in terminology provides ease in separating out use cases from varying vertical and quality domains (e.g., transportation, medical, financial, agricultural, safety-critical, security-critical, performance-critical, high assurance, to name a few). That is useful since there is no singular IoT, and it is meaningless to speak of comparing one IoT to another.

According to Voas, a NoT can be described by five primitives proposed in Voas (2016).

9.4.1.2   What Are the Five IoT Primitives?

1. Sensor: An electronic utility (e.g., proximity sensors, pressure sensors, temperature sensors, gas sensors, image sensors, acoustic sensors) that measures physical properties such as sound, weight, humidity, temperature, and acceleration. Properties of a sensor could be the transmission of data (e.g., Radio-Frequency Identification (RFID)), Internet access, and/or be able to output data based on specific events.
2. Communication channel: "a medium by which data are transmitted (e.g., physical via Universal Serial Bus, wireless, wired, and verbal)." Since data is the "blood" of a NoT, communication channels are the "veins" and "arteries", as data moves to and from intermediate events at different snapshots in time.
3. Aggregator: "a software implementation based on mathematical function(s) that transforms groups of raw data into intermediate, aggregated data. Raw data can come from any source". Aggregators have two actors for consolidating large volumes of data into lesser amounts:
   a Cluster: "an abstract grouping of sensors (along with the data they output) that can appear and disappear instantaneously".
   b Weight: "the degree to which a particular sensor's data will impact an aggregator's computation".
4. Decision trigger: "creates the final result(s) needed to satisfy the purpose, specification, and requirements of a specific NoT." decision trigger is a conditional expression that triggers an action and abstractly defines the end purpose of a NoT. A decision trigger's outputs can control actuators and transactions.
5. External utility (eUtility): "a hardware product, software, or service, which executes processes or feeds data into the overall dataflow of the NoT."

9.4.1.3   What Are the Open Issues to Consider for Software Engineering When Building IoT Systems?

There are various challenges associated with building requirements for IoT systems. First, it is a relatively new domain, and capturing the requirements based on the proper domain knowledge is necessary before designing and developing IoT-based systems. Secondly, when specifying the functionality for IoT applications, attention is naturally focused on concerns such as fitness of purpose, wireless interoperability, energy efficiency, and so on. Conventional requirements elicitations techniques such as domain analysis, joint application development (JAD), and quality function deployment (QFD) among others are usually adequate for these kinds of requirements. But in some domains, such as healthcare or education, where IoT applications can be deployed, some quality requirements are probably of greater concern.

For example, given the increased communication and complexity of IoT technology, there is an increase in security-related concerns (Georgescu & Popescu, 2015). Many of the devices used in a provisioned, specialized IoT will collect various data whether that surveillance is known or not" (Laplante, Laplante, & Voas, 2015). But why are these data being collected? Who owns the data? And where does the data go? These are questions that need to be answered by the legal profession and government entities that will oversee the deployment of IoT systems in various domains.

Conversely, by embedding sensors into front field environments as well as terminal devices, an IoT network can collect rich sensor data that reflect the real-time environment conditions in the field and the events/activities that are occurring. Since the data is collected in the granularity of elementary event level in a 7 × 24 mode, the data volume is very high and the data access pattern also differs considerably from traditional business data. The related "scalability" requirements will need to be addressed.

Finally, deploying IoT systems can uncover new quality attributes. For example, there are questions on the moral role that the IoT may play in human lives, particularly concerning personal control. Applications in the IoT involve more than computers interacting with other computers. Fundamentally, the success of the IoT will depend less on how far the technologies are connected and more on the humanization of the technologies that are connected. IoT technology may reduce people's autonomy, shift them toward particular habits, and then shift power to corporations focused on financial gain. When deploying IoT technologies in classrooms, for instance, this effectively means that the controlling agents are the organizations that control the tools used by the academic professionals but not the academic professionals themselves (Gubbi, Buyya, Marusic, & Palaniswami, 2013). The dehumanization of humans in interacting with machines is a valid concern. Many studies indicate that face-to-face interaction between students will not only benefit a child's social skills but also positively contributes toward character building. The issue that may arise from increased IoT technologies in education is the partial loss of the social aspect of going to school. Conversely, using IoT in virtual learning environments can be of special support to students with special needs, for example, dyslexic and dyscalculic needs (Lenz et al., 2016).

Similarly, when constructing IoT systems for the healthcare domain, it is important to engage all stakeholders when trying to define a notion of "caring" for a new healthcare system and it is critically important to engage systems engineers, computer scientists, doctors, nurses, and, most importantly, patients during requirements discovery. Laplante, Kassab and Laplante (2017) presented a structured approach based on the above discussed NoT primitives for describing IoT for healthcare systems while illustrating their approach for three use cases and discussing relevant quality issues that arise, in particular, the need to consider "caring" as an emerging quality requirement.

9.4.2 SOFTWARE ENGINEERING AND BLOCKCHAIN

9.4.2.1 What Is Blockchain?

The theory behind Bitcoin as a peer-to-peer electronic cash system was introduced in a white paper written under the pseudonym "Satoshi Nakamoto" in 2008 (Nakamoto, 2008). More than a decade later, and despite the uncertainty of the identity of its creator, Bitcoin was rapidly implemented and widely accepted as a prominent online cryptocurrency. This is evidenced by the total USD value of Bitcoin supply nears $1 trillion (as of March 2022). Many online retailers accept Bitcoin as a mean of payment with many mechanisms in existence for exchanging it with fiat currency and vice versa.

The blockchain is the essence of the infrastructure underlying Bitcoin and other cryptocurrencies. In practice a blockchain is built upon a chronological chain of block-like data structures, hence its name. A block hosts with a timestamped set of transactions that are bundled together. Each new block is linked to its preceding block. Combined with cryptographic hashes, this timestamped chain of blocks provides a hopefully immutable record of all transactions in a network, from the genesis block until the last/most current block.

Figure 9.1 depicts a standard structure for a blockchain, which consists of the following four pieces of metadata:

1. Previous block reference
2. Proof of work (a.k.a. a nonce)
3. Time-stamp
4. Merkle tree root for the block transactions. A blockchain comprises a set of nodes without a pre-existing trust relationship and is connected through a peer-to-peer topology.

Each node hosts the same copy of a blockchain creating a decentralized structure. However, for such a structure to be useful, there must exist some mechanism by which the nodes can mutually reach a consensus on the next valid block in the chain to be added. The consensus mechanisms are protocols that make sure all nodes (devices that maintain the blockchain and, sometimes, process transactions) are synchronized with each other and agree on which transactions are legitimate and are added to the blockchain. These consensus mechanisms are crucial for the precise function of a blockchain. Some of the schemes adopted for establishing a distributed consensus include Proof of Work, Proof of Stake, Proof of Capacity, Proof of Human-Work, Proof of Activity, and Proof of Elapsed Time.

A blockchain is an append-only distributed ledger. In other words, the new entries get added at the end of the ledger. In contrast with a traditional relational database, where data can be deleted or altered, there are no administrator's permissions within a blockchain that allow for deleting or editing of the recorded data. Furthermore, unlike a centralized relational database, blockchains are designed for decentralized applications. This immutability feature implies that once a transaction is added to the blockchain, no one can alter it. This makes blockchain an ideal solution for assets

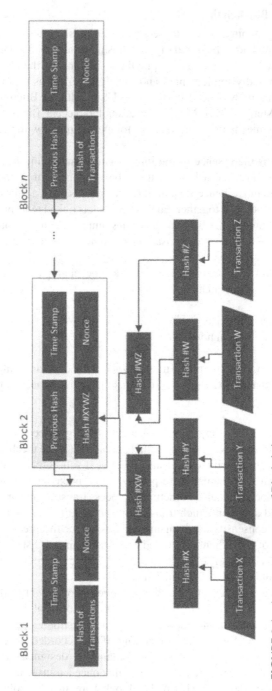

FIGURE 9.1   A common structure of Blockchain

transactions and identity management, to mention a few examples. In addition to decentralization, consensus, and immutability, a blockchain network has two additional key characteristics: Provenance and finality. Provenance refers to the awareness of participants of the network about where the asset originated from and its ownership history, while finality refers to the status of a transaction as complete.

A blockchain can use smart contracts (executable code), which are stored on the blockchain and executed automatically to serve as agreements or a set of rules that oversee a blockchain transaction. For example, a smart contract may define the contractual conditions of an individual's travel insurance. The conditions will automatically execute upon notice of a flight delay by more than a certain number of hours.

A blockchain can be both permissionless (public) and permissioned (private). In a permissionless blockchain, any node can join the network. In a permissioned (private) blockchain, pre-verification of the participating parties, which are all known to each other, is required. The choice between the two types is mainly driven by the use case in a particular application. If a network can "commoditize" trust, where the identity of the facilitating parties does not need to be verified, a permissionless blockchain makes sense. An example of a permissionless blockchain is Bitcoin or Ethereum. On the other hand, managing the medical healthcare records is an ideal use case for a permissioned blockchain because it makes sense to have the participating companies vetted. In other words, when it is vital that the blockchain participants require permission to execute transactions, a permissioned blockchain makes sense. This also helps all participants in the network to understand where the transactions are originated from. Hyperledger – an open-source blockchain initiative hosted by the Linux Foundation – is an example of a permissioned blockchain.

9.4.2.2 What Are the Capabilities of Blockchain Technology?

Blockchain technology's transformative capabilities have been rapidly recognized as a turning point in many use case scenarios beyond the financial sector. A study by Chakraborty, Shahriyar, Iqbal, and Bosu (2018) reported that 3,000 blockchain software projects were hosted on Github in March 2018. A similar search we conducted on Github in June 2021 yielded more than 90,000 projects. This emerging technology's impetus is now being utilized in multiple ways, from global payments to managing the COVID-19 vaccinations supply chain, to tracking diamond sales.

The World Economic Forum estimates 10% of the global GDP will be stored using blockchain by 2027. A renowned scientific study and market consulting firm, Gartner, estimated projected investment decisions worth $3.1 trillion in blockchain technology by 2030 (Kandaswamy & Furlonger, 2018). In synchronous with the growing sphere of blockchain, numerous data points testify to the increasing demand for blockchain developers and engineers. As of June 2021, a Web search for blockchain-related jobs in the U.S. retrieved 5,603 results on Indeed.com. The skyrocketing demand for blockchain-related jobs has also translated into a significant salary bump. According to global statistics provided by Hired.com, the median salary for a blockchain-related job opening is approximately 1.3 times higher than the standard salary for software engineers. Unlike traditional software development, blockchain engineers need to secure an immutable and decentralized database hosted on distributed nodes connected through a peer-to-peer network without a pre-existing trust relationship (Kassab, Destefanis, DeFranco, & Pranav, 2021).

9.4.2.3   What Are the Competencies Required to Become a Blockchain Engineer?

Kassab et al. (2021), reported the results from conducting a structured inspection of 400 job adverts extracted through a systematic process to analyze the essential industrial demand for competencies needed for the emerging blockchain engineer role in the U.S. job market. The emerging role of blockchain engineers is specific, requires dedicated skills, and the education sector needs to set up dedicated study paths to create these new professional figures requested by the market. Interested readers can refer to Kassab et al. (2021) for the detailed results on the required competencies.

9.4.2.4   What Are the Open Issues to Consider for Software Engineering When Building Blockchain Applications?

Although blockchain technology has the potential to disrupt traditional business models in many business domains, the interrelation between blockchain and software engineering has surprisingly received little attention and only recently (Demi, 2020). The significant differences between blockchain-oriented development and traditional software development motivated the blockchain community to propose a new development paradigm named Blockchain-Oriented Software Engineering (BOSE) (Destefanis, Marchesi, Ortu, Tonelli, Bracciali, & Hierons, 2018).

Software Engineering practices must be revisited in any blockchain-based project. For example, while there is still a significant incongruity regarding how the blockchain's unique five characteristics can be mapped to the elicited requirements (this area is open to research), Kassab (2021) presented a first attempt toward exploring NFRs for blockchain-oriented systems in particular in respect to 1) privacy, 2) performance and scalability, 3) interoperability, 4) usability, 5) compliance with regulations, and 6) operating and financial constraints. In addition, existing events such BlockArch (https://ww2.inf.ufg.br/~insight/blockarch2022/index.html) and WETSEB (www.agile-group.org/wetseb2022/) aim to foster discussions on the possible synergies between blockchain technology and software engineering practices and how both topics can be related to providing software solutions that rely on blockchain advantages. Future research agendas on the link between blockchain and software engineering include:

- Identification of the design and architectural requirements that emerge due to the five blockchain qualities. Specifically, how these qualities can be satisfied during the high-level and low-level design phases, and what are the new requirements that could be introduced due to this?
- Exploring the new emerging NFRs (not listed in the traditional qualities taxonomies) for blockchain-based systems when used as a platform for more focused domain-related applications (e.g. healthcare, education, retail, etc.)? What are these, and how they will be satisfied?

- Tools for Blockchain software distributed development and community management
- Security and reliability in Blockchain and Smart Contracts
- Agile and Lean processes for Blockchain software development.

9.4.3 SOFTWARE ENGINEERING AND ARTIFICIAL INTELLIGENCE

9.4.3.1 What Is AI?

McCarthy (2007) defined AI as the science and engineering of developing computer programs that, when fed into a machine, make the machine exhibit the intelligence of humans. While there are other definitions of AI, there is no universally accepted one. Most people, however, understand and accept the concept of AI as any system that can make decisions and perform tasks as an intelligent human would. This means that consumers and other stakeholders expect AI-based software systems to perform at least as well as humans, if not better.

9.4.3.2 Can You Give Some Examples of Areas of Research for Software Engineering for AI Systems?

AI has been well researched for use in software engineering. For example, over the years, many AI techniques have been employed to capture, represent, and analyze requirements.

Requirements elicitation is likely the most investigated requirements engineering task. One prominently explored task within elicitation is ambiguity detection, which is typically supported via a combination of natural language processing and machine learning (e.g., Chechik, 2019; Ferrari & Esuli, 2019). Another sample AI-elicitation approach is the work of Peclat and Ramos (2018) who examined using semantic analysis for identifying security concerns from collections of unstructured textual documents. The automated analysis of user feedback and emotions (part of affective computing) to capture requirements and feedback directly and automatically from the end is also gaining traction (Groen et al., 2017).

Another topic that received great attention is using AI for requirements traceability. For example, Hayes, Payne, and Leppelmeier (2019) presented the use of metadata, such as readability indexes, as a resource for requirements traceability.

Automated classification remains also one of the most mature AI techniques applied to tackle requirements engineering problems. Stanik, Haering, & Maalej (2019) presented an application for classifying reviews to better make sense and use of the user opinions shared in social media. They compared traditional machine learning with deep learning when classifying app reviews and users' tweets into problem reports (potential bugs), inquiries (feature requests), and being irrelevant (noisy feedback). Rashwan (2012) also presented a semantic analysis approach to classifying requirements into functional and non-functional requirements.

Requirements analysis practices have also been approached with AI. For example, del Sagrado and del Aguila (2018) presented an approach using Bayesian network requisites to predict whether the requirements specification documents have to be revised. Requisites provide an estimation of the degree of revision for a given requirements specification (i.e., SRS). Thus, it helps when identifying whether a requirements specification is sufficiently stable and needs no further revision.

While AI has the potential to revolutionize requirements engineering practices, many challenges still exist. Deep learning is data-hungry, which calls for a community effort in data sharing, curation, and provenance. More work on understanding the data is necessary. Also, as with other disruptive technologies, emerging quality requirements need to be fully understood

It is also worth mentioning that there is a distinction between "AI in RE" and "RE in AI". While the first focuses on how to develop machine learning and deep learning to improve requirements engineering tasks, the second focuses on how requirements engineering will contribute to AI technology. Dalpiaz, and Niu (2020) provided a summary of the current research trends in each of these two directions.

9.4.3.3   What Are the Open Issues to Consider for Software Engineering When Building AI-Based Systems?

As with other disruptive technologies, emerging quality requirements need to be fully understood. In the context of AI, there is a need to be aware of new quality requirements, such as explainability (that the actions the system take can be explained to an observer), freedom from bias (that the system does not act in a way that disadvantages any user group), and specific legal requirements (particularly those regulating the deployment of AI in safety-critical contexts).

9.4.4   SOFTWARE ENGINEERING AND CLOUD COMPUTING

9.4.4.1   What Is Cloud Computing?

In industrial practice, cloud computing is becoming increasingly used as an option for increasing productivity, reducing labor and infrastructure costs, and improving agility. According NIST, cloud computing is defined as: "a model for enabling ubiquitous, convenient, on-demand network access to a shared pool of configurable computing resources (e.g., networks, servers, storage, applications, and services) that can be rapidly provisioned and released with minimal management effort or service provider interaction" (Mell & Grance, 2011).

9.4.4.2   What Are the Open Issues to Consider for Software Engineering When Building Cloud Computing Systems?

Despite the obvious benefits, cloud-based solutions face questions relating to the best architectures, privacy concerns, difficulty complying with regulations, performance problems, availability issues. Wind and Schrödl (2010) examined selected requirements engineering approaches to study their extent to accommodate specific requirements of cloud-based solutions. They recommend that the following characteristics need to be examined when selecting a requirements engineering approach for cloud computing systems:

- The architectural capacity of a requirements engineering approach to describe the connecting element between the individual application components

- The agility of the requirements engineering approach in relation to the description of architecture such as the integration of multi-discipline components from different domains, or different requirements sources, affects the requirements engineering process
- The structured elicitation of infrastructure requirements. These infrastructure requirements must be allocated into areas of service quality, security, and economic dimension.
- Comprehensive inclusion of the customer into the entire development process during every phase. Even where this can be difficult, due to different language biases and differing levels of understanding by developers, this must not be abandoned.
- The preparation for an optimally functioning change management system for the phase following delivery, in order to be able to implement any modifications in the service area.

9.4.5 SOFTWARE ENGINEERING AND AFFECTIVE COMPUTING

9.4.5.1 What Is Affective Computing?

Affective computing is an interdisciplinary field spanning computer science, psychology, and cognitive science; it focuses on the study and development of systems that can recognize, interpret, process, and simulate human effects. Awareness of emotions to be experienced by end users is of utmost importance in requirements engineering and therefore emotional requirements should be always considered. These emotions and stakeholders' cognitive states arise and evolve differently during the elicitations, prioritization, and negotiation processes. Hence, it is becoming increasingly important to understand these emotions and effects to be experienced by stakeholders within sociotechnical systems. The sentiment expressed together with opinions helps to assess important topics and creates actionable insights.

9.4.5.2 What Are the Open Issues to Consider for Software Engineering When Building Affective Computing Systems?

Trending research topics to further explore the relation between affective computing in software engineering include:

- Methods and artifacts for elicitation and modeling of emotional requirements, including the relevant approaches of participatory requirements engineering
- The potential and challenges of different types of personality characteristics of software engineers when conducting requirements engineers along with the limitation regarding the current state of the art
- Defining or adapting psychological models of affect to RE (e.g., understanding what may trigger positive or negative emotions during the requirement engineering process)
- Exploration of biometric sensors emerging from hardware (e.g., smartwatch) enables new measurement techniques to support the V&V of requirements.

NOTE

1   Much or this chapter is adapted from Phillip A. Laplante and Mohamad H. Kassab, *Requirements Engineering for Software and Systems, Fourth Edition*, Taylor & Francis, 2022, with permission.

FURTHER READING

Azar, J., Smith, R.K., & Cordes, D. (2007). Value-oriented requirements prioritization in a small development organization. *IEEE Software*, 24(1), 32–37.

Bhat, J.M., Mayank G., & Santhosh N.M. (2006) Overcoming requirements engineering challenges: Lessons from offshore outsourcing. *IEEE software 23.5*: 38–44.

Bose, S., Kurhekar, M., & Ghoshal, J. (2008). Agile methodology in Requirements Engineering. *SETLabs Briefings Online*, 13–21.

Chakraborty, P., Shahriyar, R., Iqbal, A., & Bosu, A. (2018, October). Understanding the software development practices of blockchain projects: A survey. In *Proceedings of the 12th ACM/IEEE International Symposium on Empirical Software Engineering and Measurement* (pp. 1–10).

Chechik, M. (2019, September). Uncertain requirements, assurance and machine learning. In *2019 IEEE 27th International Requirements Engineering Conference (RE)* (pp. 2–3). IEEE.

Dalpiaz, F., & Niu, N. (2020). Requirements engineering in the days of artificial intelligence. *IEEE Software*, 37(4), 7–10.

del Sagrado, J., & del Aguila, I.M. (2018). Stability prediction of the software requirements specification. *Software Quality Journal*, 26(2), 585–605.

Demi, S. (2020, August). Blockchain-oriented requirements engineering: A framework. In *2020 IEEE 28th International Requirements Engineering Conference (RE)* (pp. 428–433). IEEE.

Destefanis, G., Marchesi, M., Ortu, M., Tonelli, R., Bracciali, A., & Hierons, R. (2018, March). Smart contracts vulnerabilities: a call for blockchain software engineering? In *2018 International Workshop on Blockchain Oriented Software Engineering (IWBOSE)* (pp. 19–25). IEEE.

European Commission (2005). *The new SME definition: user guide and model declaration section*. Office for Official Publications of the European Communities, Brussels.

Ferrari, A., & Esuli, A. (2019). An NLP approach for cross-domain ambiguity detection in requirements engineering. *Automated Software Engineering*, 26(3), 559–598.

Georgescu, M., & Popescu, D. (2015). How could internet of things change the E-learning environment. The 11th International Scientific Conference eLearning and Software for Education.

Grama, J. (2014). *Just in time research: Data breaches in higher education*. EDUCAUSE.

Groen, E.C., Seyff, N., Ali, R., Dalpiaz, F., Doerr, J., Guzman, E., ... & Stade, M. (2017). The crowd in requirements engineering: The landscape and challenges. *IEEE software*, 34(2), 44–52.

Gubbi, J., Buyya, R., Marusic, S., & Palaniswami, M. 2013. Internet of Things (IoT): A vision, architectural elements, and future directions. *Future generation computer systems*, 29(7), 1645–1660.

Han, W. (2011). Research of intelligent campus system based on IOT. In *Advances in Multimedia, Software Engineering and Computing*, 1, Springer, 165–169.

Hayes, J.H., Payne, J., & Leppelmeier, M. (2019, September). Toward Improved Artificial Intelligence in Requirements Engineering: Metadata for Tracing Datasets. In *2019 IEEE*

27th International Requirements Engineering Conference Workshops (REW) (pp. 256–262). IEEE.

Hofstede, G. (2001). Culture's Consequences: Comparing Values, Behaviors, Institutions and Organizations Across Nations. Sage Publications.

JPMorgan Chase Co. (2018). Small businesses are an anchor of the us economy. www.jpmorganchase.com/institute/research/smallbusiness/small-business-dashboard/economic-activity.

Kandaswamy, R., & Furlonger, D. (2018). Blockchain-based transformation: A Gartner trend insight report.

Kassab, M. (2015, August). The changing landscape of requirements engineering practices over the past decade. In 2015 IEEE Fifth international workshop on empirical requirements engineering (EmpiRE) (pp. 1–8). IEEE.

Kassab, M. (2021). How Requirements Engineering is Performed in Small Businesses? Proceeding of the Workshop on Requirement Engineering for Software startups and Emerging Technologies in conjunction with 29th IEEE International Requirements Engineering Conference 2021.

Kassab, M., Destefanis, G., DeFranco, J., & Pranav, P. (2021, May). Blockchain-Engineers Wanted: an Empirical Analysis on Required Skills, Education and Experience. In 2021 IEEE/ACM 4th International Workshop on Emerging Trends in Software Engineering for Blockchain (WETSEB) (pp. 49–55). IEEE.

Laplante, P.A., Kassab, M., Laplante, N.L., & Voas, J.M. (2017). Building caring healthcare systems in the Internet of Things. IEEE Systems Journal, 12(3), 3030–3037.

Laplante, P.A., Laplante, N., & Voas, J. (2015). Considerations for healthcare applications in the internet of things. Rel. Dig., 61(4), 8–9.

Lenz, L., Pomp, A., Meisen, T., & Jeschke, S. (2016). How will the Internet of Things and Big Data analytics impact the education of learning-disabled students? A concept paper. In 3rd MEC International Conference on Big Data and Smart City (ICBDSC) (pp. 1–7).

McCarthy, J. (2007). What Is Artificial Intelligence? Computer Science Department, Stanford University, Stanford, CA 94305

Mell, P., & Grance, T. (2011). The NIST definition of cloud computing. Available at: https://csrc.nist.gov/publications/detail/sp/800-145/final

Nakamoto, S. (2008). Bitcoin: A peer-to-peer electronic cash system. Decentralized Business Review, 21260.

Niazi, M., Mahmood, S., Alshayeb, M., Riaz, M.R., Faisal, K., Cerpa, N., ... & Richardson, I. (2016). Challenges of project management in global software development: A client-vendor analysis. Information and Software Technology, 80, 1–19. www-formal.stanford.edu.ezaccess.libraries.psu.edu/jmc/whatisai.pdf

Peclat, R.N., & Ramos, G.N. (2018). Semantic analysis for identifying security concerns in software procurement edicts. New Generation Computing, 36(1), 21–40.

Rashwan, A. (2012, May). Semantic analysis of functional and non-functional requirements in software requirements specifications. In Canadian Conference on Artificial Intelligence (pp. 388–391). Springer, Berlin, Heidelberg.

Sangwan, R.S., & Laplante, P.A. (2006). Test-driven development in large projects. IT Professional 8.5: 25–29.

Satzger, B., Zabolotnyi, R., Dustdar, S., Wild, S., Gaedke, M., Göbel, S., & Nestler, T. (2014). Toward collaborative software engineering leveraging the crowd. In Economics-Driven Software Architecture (pp. 159–182). Morgan Kaufmann.

Schmid, K. (2014). Challenges and Solutions in Global Requirements Engineering – A Literature Survey. In: Winkler D., Biffl S., Bergsmann J. (eds) Software Quality. Model-Based

Approaches for Advanced Software and Systems Engineering. SWQD 2014. Lecture Notes in Business Information Processing, vol 166. Springer, Cham. https://doi.org/ 10.1007/978-3-319-03602-1_6

Sinha, V., & Sengupta, B. (2006). Enabling collaboration in distributed requirements management. *Software, 23:* 52–61.

Stanik, C., Haering, M., & Maalej, W. (2019, September). Classifying multilingual user feedback using traditional machine learning and deep learning. In *2019 IEEE 27th International Requirements Engineering Conference Workshops (REW)* (pp. 220–226). IEEE.

Sula, A., Spaho, E., Matsuo, K., Barolli, L., Miho, R., & Xhafa, F. (2013). An IoT-based system for supporting children with autism spectrum disorder. In *2013 Eighth International Conference on Broadband and Wireless Computing, Communication and Applications*, pp. 282–289.

U.S. Small Business Administration. (2021). www.sba.gov/

Uzelac, A., Gligoric, N., & Krco, S. (2015). A comprehensive study of parameters in physical environment that impact students' focus during lecture using Internet of Things. *Computers in Human Behavior, 53*, 427–434.

Voas, J.M. (2016). NIST SP 800-183 Networks of 'Things': http://dx.doi.org/10.6028/NIST. SP.800-183 http://nvlpubs.nist.gov/nistpubs/SpecialPublications/NIST.SP.800-183.pdf

Wind, S., & Schrödl, H. (2010, December). Requirements engineering for cloud computing: a comparison framework. In *International Conference on Web Information Systems Engineering* (pp. 404–415). Springer, Berlin, Heidelberg.

Appendix A
Software Requirements for a Wastewater Pumping Station Wet Well Control System (rev. 02.01.00)

Christopher M. Garrell

A.1 INTRODUCTION

A wastewater pumping station is a component of the sanitary sewage collection system that transfers domestic sewage to a wastewater treatment facility for processing. A typical pumping station includes three components: (1) a sewage grinder, (2) a wet well, and (3) a valve vault (Figure A.1). Unprocessed sewage enters the sewage grinder unit so that solids suspended in the liquid can be reduced in size by a central cutting stack. The processed liquid then proceeds to the wet well, which serves as a reservoir for submersible pumps. These pumps then add the required energy/head to the liquid so that it can be conveyed to a wastewater treatment facility for primary and secondary treatment. The control system specification that follows describes the operation of the wet well.

A.1.1 PURPOSE

This specification describes the software design requirements for the wet well control system of a wastewater pumping station. It is intended that this specification provide the basis of the software development process and be used as preliminary documentation for end users.

A.1.2 SCOPE

The software system described in this specification is part of a control system for the wet well of a wastewater pumping station. The control system supports an array of sensors and switches that monitor and control the operation of the wet well. The design of the wet well control system shall provide for the safety and protection of pumping station operators, maintenance personnel, and the public from hazards

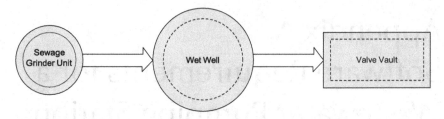

FIGURE A.1   Typical wastewater pumping station process

that may result from its operation. The control system shall be responsible for the following operations:

1. Monitoring and reporting the level of liquid in the wet well
2. Monitoring and reporting the level of hazardous methane gas
3. Monitoring and reporting the state of each pump and noting whether it is currently running or not
4. Activating a visual and audible alarm when a hazardous condition exists
5. Switching each submersible pump on or off in a timely fashion depending on the level of liquid within the wet well
6. Switching ventilation fans on or off in a timely fashion depends on the concentration of hazardous gas within the wet well.

Any requirements that are incomplete are annotated with "TBD" and will be completed in a later revision of this specification.

A.1.3   DEFINITIONS, ACRONYMS, AND ABBREVIATIONS

The following is a list of definitions for terms used in this document.

audible alarm	The horn that sounds when an alarm condition occurs.
controller	Equipment or a program within a control system that responds to changes in a measured value by initiating an action to affect that value.
DEP	Department of Environmental Protection.
detention basin	A storage site, such as a small, unregulated reservoir, which delays the conveyance of wastewater.
effluent	Any material that flows outward from something; an example is a wastewater from treatment plants.
EPA	Environmental Protection Agency.
imminent threat	A situation with the potential to immediately and adversely affect or threaten public health or safety.
influent	Any material that flows inward from something; an example is a wastewater into treatment plants.
manhole	Hole, with removable cover, through which a person can enter into a sewer, conduit, or tunnel to repair or inspect.

methane	A gas formed naturally by the decomposition of organic matter.
overflow	An occurrence by which a surplus of liquid exceeds the limit or capacity of the well.
pre-cast	A concrete unit that is cast and cured in an area other than its final position or place.
pump	A mechanical device that transports fluid by pressure or suction.
remote override	A software interface that allows remote administrative control of the pumping control system.
seal	A device mounted in the pump housing and/or on the pump shaft, to prevent leakage of liquid from the pump.
security	Means used to protect against unauthorized access or dangerous conditions; a resultant visual and/or audible alarm is then triggered.
sensor	The part of a measuring instrument that responds directly to changes in the environment.
sewage grinder	A mechanism that captures, grinds, and removes solids, ensuring a uniform particle size to protect pumps from clogging.
submersible pump	A pump having a sealed motor that is submerged in the fluid to be pumped.
thermal overload	A state in which measured temperatures have exceeded a maximum allowable design value.
valve	A mechanical device for controlling the flow of a fluid.
ventilation	The process of supplying or removing air by natural or mechanical means to or from space.
visible alarm	The strobe light that is enabled when an alarm condition occurs.
voltage	Electrical potential or electromotive force expressed in volts.
wet well	A tank or separate compartment following the sewage grinder that serves as a reservoir for the submersible pump.

A.2   OVERALL DESCRIPTION

A.2.1   Wet Well Overview

The wet well for which this specification is intended is shown in Figure A.2. The characteristics of the wet well described in this specification are as follows.

1. The wet well reservoir contains two submersible pumps sized to provide a fixed capacity.
2. Hazardous concentrations of flammable gases and vapors can exist in the wet well.
3. It has a ventilation fan that is oriented to direct fresh air into the wet well rather than just removing the exhaust from the well.
4. An alarm and indicator light are located outside so that operators can determine if a hazardous condition exists. Hazardous conditions include, but are

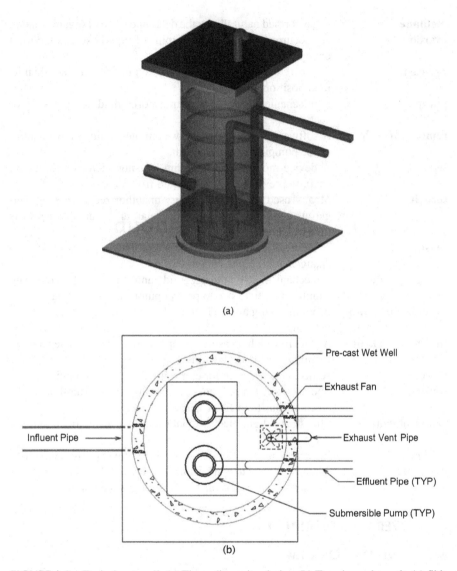

FIGURE A.2 Typical wet well: (a) Three-dimensional view (b) Top view schematic (c) Side sectional view

not necessarily limited to, a high gas level, a high water level, and pump malfunction.

5. A float switch is used to determine the depth of liquid currently in the wet well.

A.2.2 PRODUCT PERSPECTIVE

A.2.2.1 System Interfaces

The system interfaces are described below.

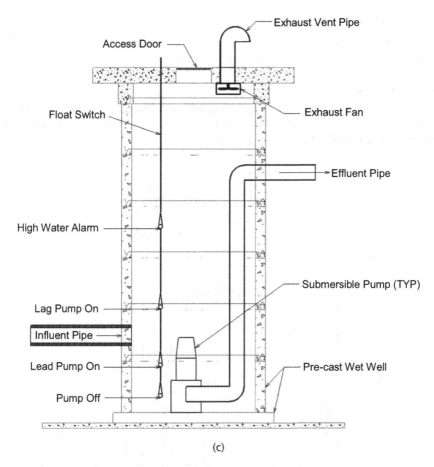

(c)

FIGURE A.2   (Continued)

A.2.2.2   User Interfaces

Pumping Station Operator
The pumping station operator uses the control display panel and alarm display panel to control and observe the operation of the submersible pumps and wet well environmental conditions. Manipulation of parameters and the state of the submersible pumps is available when the system is running in manual mode.

Maintenance Personnel
The maintenance personnel use the control display panel and alarm display panel to observe the current parameters and state of the submersible pumps and wet well and perform maintenance.

A.2.2.3   Hardware Interfaces

The wet well control system hardware interfaces are summarized in Figure A.3. The major hardware components are summarized in Table A.1.

Moisture Sensor

Each submersible pump shall be equipped with a moisture sensor that detects the occurrence of an external pump seal failure. Should a seal failure be detected, the pump shall be turned off and the alarm state set.

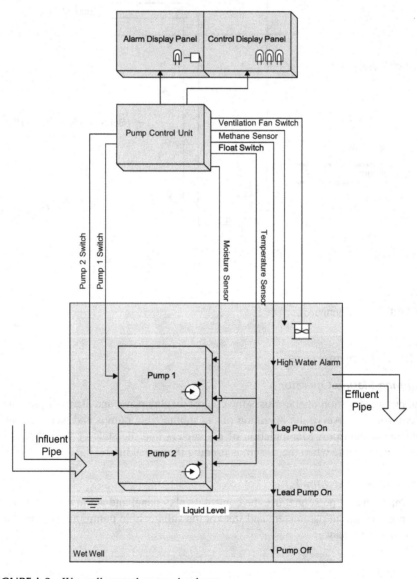

FIGURE A.3   Wet well control system hardware

TABLE A.1
Major Wet Well Control System Hardware Components

Item	Description	Quantity
1	Pre-cast concrete wet well	1
2	Access door	1
3	Ventilation pipe	2
4	Axial flow fan	2
4.1	Fan switch	2
5	Submersible pump	2
6	Pump control unit	1
6.1	Temperature sensor	2
6.2	Moisture sensor	2
6.3	Float switch	1
6.4	Access door sensor	1
7	Alarm panel	1
7.1	Alarm lamp	1
7.2	Alarm buzzer	1
8	Control panel	1
8.1	Panel lamps	6 (3 per pump)

Float Switch

The float switch is a mercury switch used to determine the depth of liquid within the wet well and set the on or off state for each pump. Three switch states have been identified as the lead pump on/off, lag pump on/off, and high water alarm.

Access Door Sensor

The access door sensor is used to determine the state, either opened or closed, of the wet well access door.

A.2.2.4   Software Interfaces

Pump Control Unit

The wet well control system interfaces with the pump control system, providing a pump station operator and maintenance personnel with the ability to observe the operation of the submersible pumps and wet well environmental conditions. The pump control unit provides the additional capability of manipulation of parameters and states of the submersible pumps when the system is running in manual mode.

Control Display Panel

The control display panel interfaces with the pump control unit, providing visual information relating to the operation of the submersible pumps and environmental conditions within the wet well.

Alarm Display Panel

The alarm display panel interfaces with the pump control unit, providing visual and audible information relating to the operation of the submersible pumps and the environmental conditions within the wet well.

A.2.2.5 Operations

The wet well control system shall provide the following operations:

1. Automated operation
2. Local manual override operation
3. Local observational operation.

A.2.3 PRODUCT FUNCTIONS

The wet well control system shall provide the following functionality.

1. Start the pump motors to prevent the wet well from running over and stop the pump motors before the wet well runs dry
2. Keep track of whether or not each motor is running
3. Monitor the pumping site for unauthorized entry or trespass
4. Monitor the environmental conditions within the wet well
5. Monitor the physical condition of each pump for the existence of moisture and excessive temperatures
6. Display real-time and historical operational parameters
7. Provide an alarm feature
8. Provide a manual override of the site
9. Provide automated operation of the site
10. Equalize the run time between the pumps.

A.2.4 USER CHARACTERISTICS

maintenance personnel   Authorized personnel trained in the usage of the wet well control system.

pumping station operator   Authorized personnel trained with the usage of the wet well control system when it is in manual mode.

A.2.5 CONSTRAINTS

System constraints include the following items:

1. Regulatory agencies include but are not limited to the Environmental Protection Agency and DEP
2. Hardware limitations
3. Interfaces to other applications
4. Security considerations
5. Safety considerations.

A.2.6 ASSUMPTIONS AND DEPENDENCIES

Assumptions and dependencies for the wet well control system include the following items:

1. The operation of the sewage grinder unit is within expected tolerances and constraints at all times.
2. A power backup system has been provided as a separate system external to the wet well control system.
3. The operation of the controls within the valve vault is within expected tolerances at all times.

A.3 SPECIFIC REQUIREMENTS

The following section defines the basic functionality of the wet well control system.

A.3.1 EXTERNAL INTERFACE REQUIREMENTS

See Figure A.4 for the use case diagram.

A.3.2 CLASSES/OBJECTS

A.3.2.1 Pump Control Unit

The pump control unit shall start the submersible pump motors to prevent the wet well from running over and stop the pump motors before the wet well runs dry within

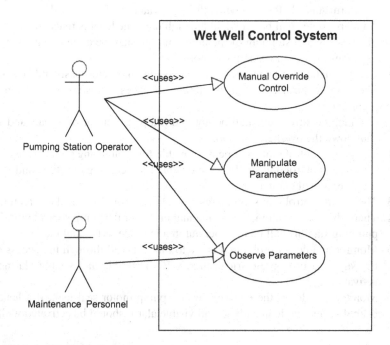

FIGURE A.4 Use case diagram

5 seconds. LeadDepth represents the depth of liquid when the first pump should be turned on. LagDepth represents the depth of liquid when the second pump should be turned on. HighDepth represents the depth of liquid that the wet well should be kept below. Should the depth of liquid be equal to or exceed HighDepth, the alarm state is set. AlarmState represents a Boolean quantity such that at any time t, the audible and visual alarms are enabled. Depth represents the amount of liquid in the wet well at any time t in units of length. Pumping represents a Boolean quantity such that at any time t, the pumps are either running or not.

$Depth: Time \rightarrow Length$
$Pumping: Time \rightarrow Bool$
$HighDepth > LagDepth > LeadDepth$
$Depth \geq LagDepth \Rightarrow Pumping$
$Depth \geq HighDepth \Rightarrow AlarmState$

1. The pump control unit shall start the ventilation fans in the wet well to prevent the introduction of methane into the wet well within 5 seconds of detecting a high methane level.
2. The pump control unit shall keep track of whether or not each motor is running.
3. The pump control unit shall keep track of whether or not each motor is available to run.
4. If a pump motor is not available to run and a request has been made for the pump motor to start, an alternative motor should be started in its place.
5. An alarm state shall be set when the high water level is reached.
6. An alarm state shall be set when the high methane level is reached.
7. The starting and stopping of the pump motors shall be done in a manner that equalizes the run times on the motors.
8. Level switches shall be used to indicate when pump motors should be started.
9. The pump control unit shall be notified if excess moisture is detected in a pump motor.
10. The pump control unit shall be notified if a pump motor overheats and shall shut down the overheated motor.
11. The pump control unit shall be responsible for monitoring the pumping site.
12. The pump control unit shall be responsible for recording real-time and historical operational parameters.
13. The pump control unit shall be responsible for providing an alarm feature.
14. There shall be an automatic and manual mode for the pump control unit. Each pumping station shall be in either automatic mode or manual mode.
15. Monitor and detect prohibited entry to the wet well through the access door by way of a broken electrical circuit. Both audible and visible alarms are activated.
16. Monitor and detect the occurrence of a pump motor seal leak. If a leak has been detected, both an audible and visible alarm should be activated within 5 seconds.

A.3.2.2   Control Display Panel

1. The control display panel shall have a digital depth of influent measured in feet.
2. Monitoring and detection of prohibited entry by way of a broken electrical circuit shall be provided. Both audible and visible alarms are activated.
3. The pump control unit shall be responsible for displaying real-time and historical operational parameters.
4. Indicator lights shall be provided for the pump running state.
5. Indicator lights shall be provided for the pump seal failure state.
6. Indicator lights shall be provided for the pump's high-temperature failure state.
7. Indicator lights shall be provided for a high wet well level alarm state.

A.3.2.3   Alarm Display Panel

1. Indicator lights shall be enabled when an alarm state is activated.
2. A buzzer shall sound when an alarm state is activated.

A.3.2.4   Float Switch

1. When the depth of liquid is equal to or greater than the lead pump depth, the float switch shall set a state which causes the first pump to turn on.
2. When the depth of liquid is equal to or greater than the lag pump depth, the float switch shall set a state which causes the second pump to turn on.
3. When the depth of liquid is equal to or greater than the allowable high liquid depth, the float switch shall set an alarm state.

A.3.2.5   Methane Sensor

1. When the volume of methane is equal to or greater than the high methane volume, the methane sensor shall set a state that causes the ventilation fans to turn on within 5 seconds.
2. When the volume of methane is equal to or greater than the allowable maximum methane volume, the methane sensor shall set an alarm state.
3. HighMethane represents the volume of methane that should cause the exhaust fans to turn on.
4. MaxMethane represents the volume of methane below which the wet well should be kept. Should the volume of methane be equal to or exceed MaxMethane, an alarm state is set.
5. Exhaust fan represents a Boolean quantity such that at any time t, the exhaust fan is either running or not running.
6. AlarmState represents a Boolean quantity such that at any time t, the audible and visual alarms are enabled.

$MaxMethane > HighMethane$
$ExhaustFan : Time \rightarrow Bool$
$AlarmState : Time \rightarrow Bool$
$Methane \geq MaxMethane \Rightarrow ExhaustFan$

$$Methane < MaxMethane \Rightarrow ExhaustFan$$
$$Methane \geq MaxMethane \Rightarrow AlarmState$$

REFERENCE

IEEE Recommended Practice for Software Requirements Specifications (IEEE Std. 830-1998).

Appendix B
Software Design for a Wastewater Pumping Station Wet Well Control System (rev. 02.01.00)

Christopher M. Garrell

B.1 INTRODUCTION

A wastewater pumping station is a component of the sanitary sewage collection system that transfers domestic sewage to a wastewater treatment facility for processing. A typical pumping station includes three components: (1) a sewage grinder, (2) a wet well, and (3) a valve vault (Figure B.1). Unprocessed sewage enters the sewage grinder unit so that solids suspended in the liquid can be reduced in size by a central cutting stack. The processed liquid then proceeds to the wet well, which serves as a reservoir for submersible pumps. These pumps then add the required energy/head to the liquid so that it can be conveyed to a wastewater treatment facility for primary and secondary treatment. The control system specification that follows describes the operation of the wet well.

B.1.1 PURPOSE

This specification describes the software design guidelines for the wet well control system of a wastewater pumping station. It is intended that this specification provide the basis of the software development process and is intended for use by software developers.

B.1.2 SCOPE

The software system described in this specification is part of a control system for the wet well of a wastewater pumping station. The control system supports an array of sensors and switches that monitor and control the operation of the wet well. The design of the wet well control system shall provide for the safety and protection of pumping station operators, maintenance personnel, and the public from hazards that may result from its operation. The control system shall be responsible for the following operations:

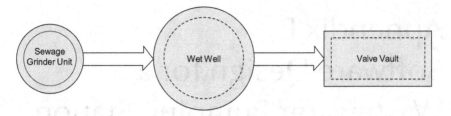

FIGURE B.1   Typical wastewater pumping station process

1. Monitoring and reporting the level of liquid in the wet well
2. Monitoring and reporting the level of hazardous methane gas
3. Monitoring and reporting the state of each pump and whether it is currently running or not
4. Activating a visual and audible alarm when a hazardous condition exists
5. Switching each submersible pump on or off in a timely fashion depending on the level of liquid within the wet well
6. Switching ventilation fans on or off in a timely fashion depending on the concentration of hazardous gas within the wet well.

Any requirements that are incomplete are annotated with "TBD" and will be completed in a later revision of this specification.

B.1.3   Definitions, Acronyms, and Abbreviations

The following is a list of definitions for terms used in this document.

attribute   Property of a class.
class   A category from which object instances are created.
message   A means of passing control from one software code unit to another software code unit because of an event.
method   A section of software code that is associated with a class providing a mechanism for accessing the data stored in the class.

B.2   OVERALL DESCRIPTION

B.2.1   Wet Well Overview

The wet well for which this specification is intended is shown in Figure B.2. This figure has been repeated from Appendix A. A more detailed description of the wet well can be found in Appendix A.

B.2.2   Wet Well Software Architecture

The wet well software architecture is shown in Figure B.3.

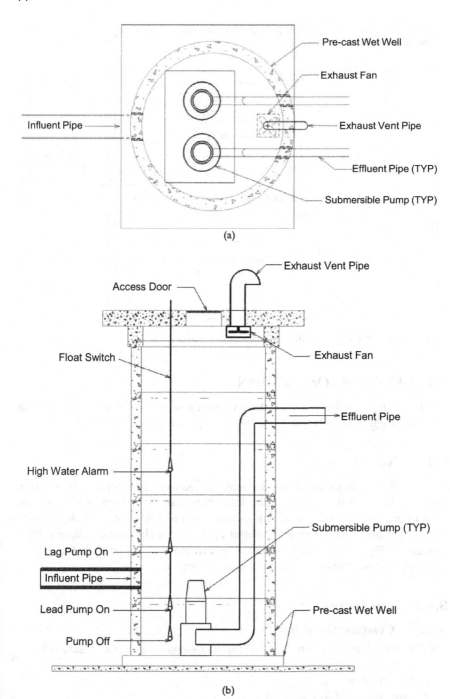

FIGURE B.2 Typical wet well. (a) Top view schematic, and (b) side sectional view

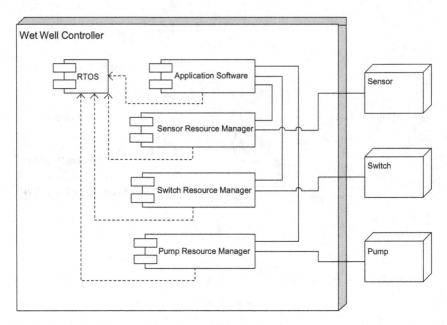

FIGURE B.3   Wet well controller software architecture

B.3   DESIGN DECOMPOSITION

The following section details the design decomposition of the wet well controller software design. This is based on the use cases presented in Appendix A.

B.3.1   CLASS MODEL

1. Figure B.4 describes the classes that make up the wet well control system software application. Figure B.5 describes the classes that make up the sensor state management of the wet well control system software application. Figure B.6 describes the classes that make up the process control of the wet well control system software application. Figure B.7 describes the classes that make up the resource logging control of the wet well control system software application.

B.3.2   CLASS DETAILS

B.3.2.1   CWetWellSimulator

The CWetWellSimulator is responsible for the following functions (Figure B.8):

1. Initialization
2. Instantiation of its contained objects
3. Monitoring and reporting the level of liquid in the wet well
4. Monitoring and reporting the level of hazardous methane gas

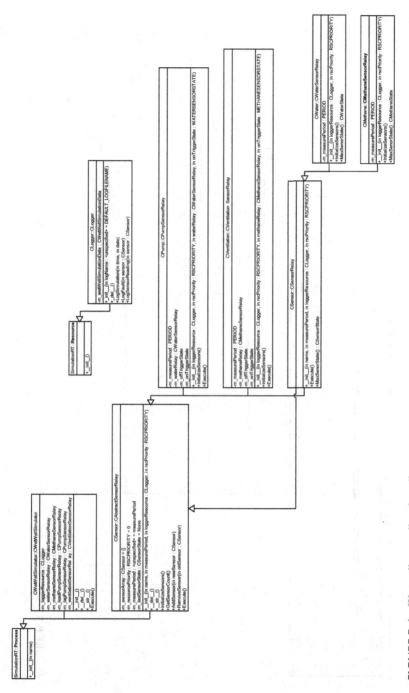

FIGURE B.4 Wet well controller class diagram

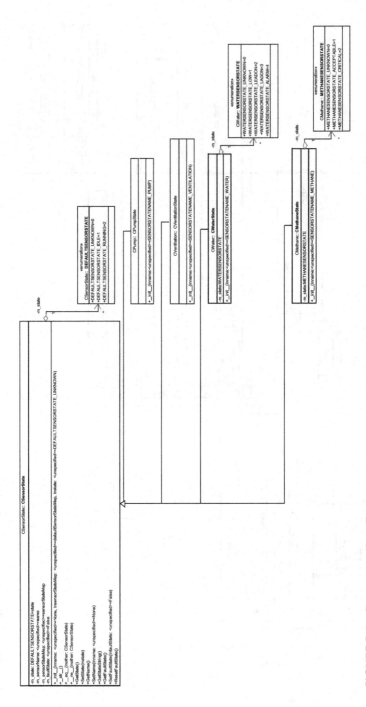

FIGURE B.5 Sensor state class diagram

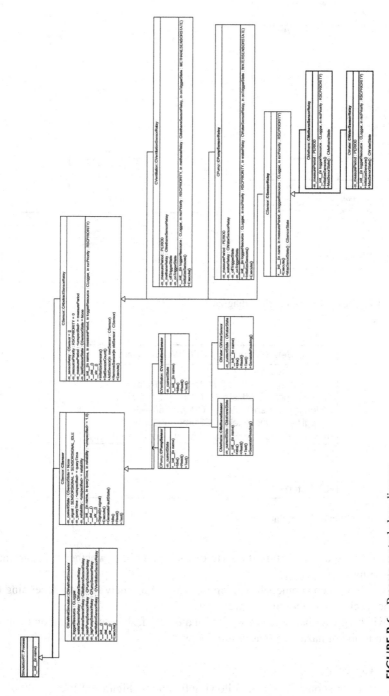

FIGURE B.6 Process control class diagram

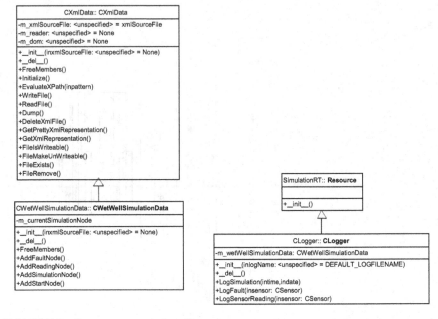

FIGURE B.7 Resource control class diagram

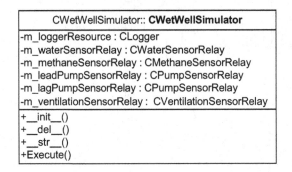

FIGURE B.8 CWetWellSimulator class

5. Monitoring and reporting the state of each pump and whether it is currently running
6. Switching each submersible pump on or off in a timely fashion depending on the level of liquid within the wet well
7. Switching ventilation fans on or off in a timely fashion depending on the concentration of hazardous gas within the wet well.

B.3.2.2   CLogger

The CLogger is responsible for the following functions (Figure B.9):

1. Initializing the simulation data logging XML file
2. Managing the logging resource mechanism for the wet well and its sensors
3. Logging each time the wet well control system is instantiated
4. Logging sensor faults
5. Logging sensor readings.

B.3.2.3 CXmlData

The CXmlData is responsible for the following functions (Figure B.10):

1. Managing generic XML content
2. Controlling XML file read and write access
3. Adding XML elements to an XML file
4. Adding XML attributes to an XML file
5. Traversing XML nodes.

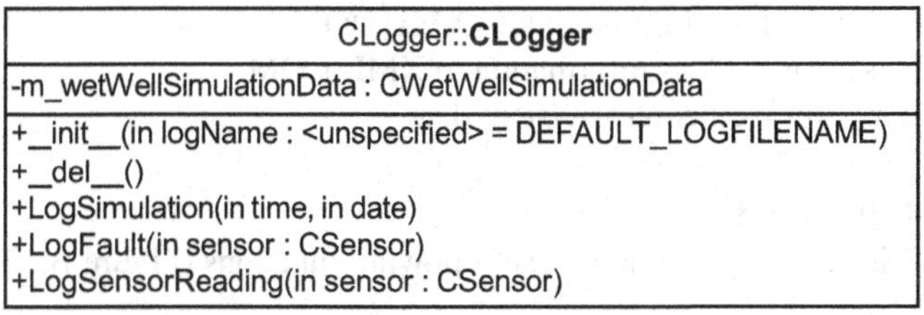

CLogger::**CLogger**
-m_wetWellSimulationData : CWetWellSimulationData
+_init_(in logName : \<unspecified\> = DEFAULT_LOGFILENAME) +_del_() +LogSimulation(in time, in date) +LogFault(in sensor : CSensor) +LogSensorReading(in sensor : CSensor)

FIGURE B.9 CLogger class

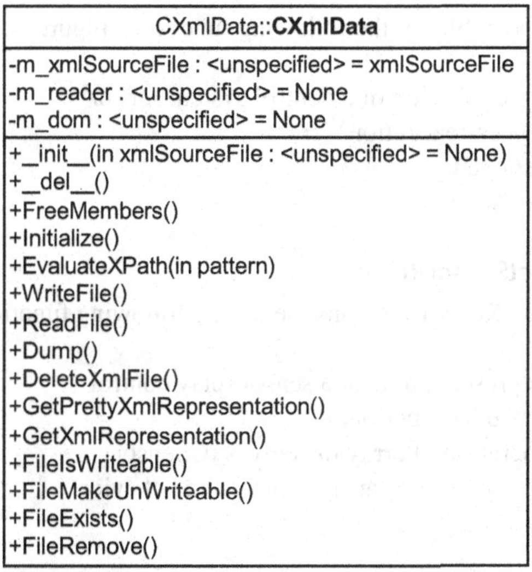

CXmlData::**CXmlData**
-m_xmlSourceFile : \<unspecified\> = xmlSourceFile -m_reader : \<unspecified\> = None -m_dom : \<unspecified\> = None
+_init_(in xmlSourceFile : \<unspecified\> = None) +_del_() +FreeMembers() +Initialize() +EvaluateXPath(in pattern) +WriteFile() +ReadFile() +Dump() +DeleteXmlFile() +GetPrettyXmlRepresentation() +GetXmlRepresentation() +FileIsWriteable() +FileMakeUnWriteable() +FileExists() +FileRemove()

FIGURE B.10 CXmlData class

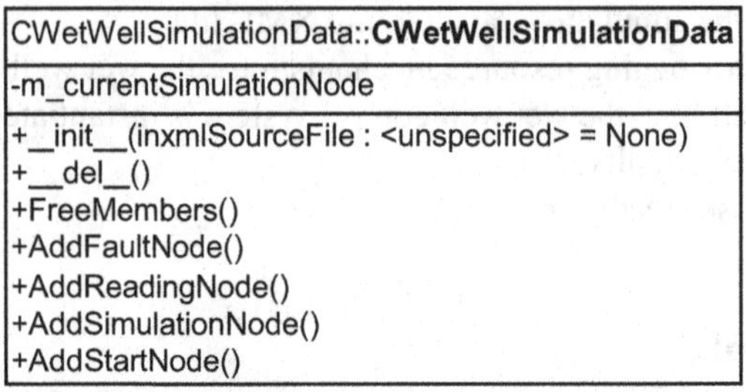

CWetWellSimulationData::**CWetWellSimulationData**
-m_currentSimulationNode
+__init__(inxmlSourceFile : \<unspecified\> = None) +__del__() +FreeMembers() +AddFaultNode() +AddReadingNode() +AddSimulationNode() +AddStartNode()

FIGURE B.11   CWetWellSimulationData class

B.3.2.4   CWetWellSimulationData

The CWetWellSimulationData is responsible for the following functions (Figure B.11):

1. Representing wet well control system operation in an XML format
2. Adding sensor fault information to XML DOM
3. Adding sensor reading information to XML DOM
4. Managing wet well control system data.

B.3.2.5   CSensorState

The CSensorState is responsible for the following functions (Figure B.12):

1. Maintaining the operational state of a sensor
2. Maintaining the fault state of a sensor.

B.3.2.6   CSensor

The CSensor is responsible for the following functions (Figure B.13):

1. A process representation of a control system sensor
2. Managing process execution
3. Reading sensor state
4. Storing sensor state.

B.3.2.7   CAbstractSensorRelay

The CAbstractSensorRelay is responsible for the following functions (Figure B.14):

1. Processing representations of a sensor relay control
2. Managing process execution
3. Managing operation of array of sensors (CSensor)
4. Providing abstract sensor array control and behavior.

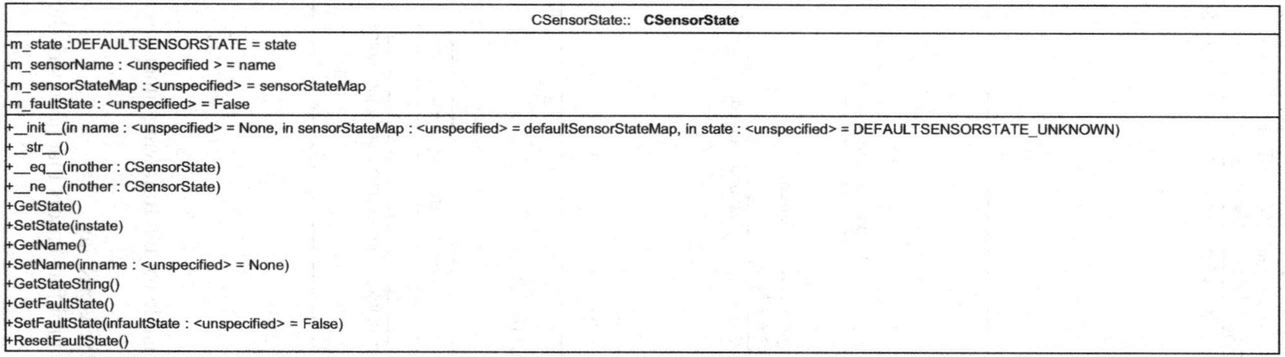

CSensorState:: **CSensorState**
-m_state :DEFAULTSENSORSTATE = state
-m_sensorName : <unspecified > = name
-m_sensorStateMap : <unspecified> = sensorStateMap
-m_faultState : <unspecified> = False
+__init__(in name : <unspecified> = None, in sensorStateMap : <unspecified> = defaultSensorStateMap, in state : <unspecified> = DEFAULTSENSORSTATE_UNKNOWN)
+__str__()
+__eq__(inother : CSensorState)
+__ne__(inother : CSensorState)
+GetState()
+SetState(instate)
+GetName()
+SetName(inname : <unspecified> = None)
+GetStateString()
+GetFaultState()
+SetFaultState(infaultState : <unspecified> = False)
+ResetFaultState()

FIGURE B.12 CSensorState class (© 2007 by Taylor & Francis Group, LLC)

CSensor::**CSensor**
-m_currentState : CSensorState = None -m_signal : SENSORSIGNAL = SENSORSIGNAL_IDLE -m_queryTime : \<unspecified\> = queryTime -m_reliability : \<unspecified\> = reliability
+__init__(in name, in queryTime, in reliability : \<unspecified\> = 1.0) +__del__() +__str__() +Signal(in signal) +Execute() +GenerateFaultState() +Idle() +Read() +Test()

FIGURE B.13   CSensor class

CSensor::**CAbstractSensorRelay**
-m_sensorArray : CSensor = [] -m_resourcePriority : RSCPRIORITY = 0 -m_measurePeriod : \<unspecified\> = measurePeriod -m_lastSensorState : CSensorState = None
+__init__(in name, in measurePeriod, in loggerResource : CLogger, in rscPriority : RSCPRIORITY) +__del__() +__str__() +InitializeSensors() +GetSensorCount() +AddSensor(in newSensor : CSensor) +RemoveSensor(in oldSensor : CSensor) +Execute()

FIGURE B.14   CAbstractSensorRelay class

CSensor::**CSensorRelay**
+__init__(inname, inmeasurePeriod, inloggerResource : CLogger, inrscPriority : RSCPRIORITY) +Execute() +MaxSenorState():CSensorState

FIGURE B.15   CSensorRelay class

B.3.2.8   CSensorRelay

The CSensorRelay is responsible for the following functions (Figure B.15):

1. Extending CAbstractSensorRelay
2. Providing process control for sensors that take periodic reading.

B.3.2.9   CMethaneState

The CMethaneState is responsible for the following functions (Figure B.16):

1. Extending CSensorState
2. Maintaining the operational state of a methane level sensor
3. Maintaining the fault state of a methane level sensor.

B.3.2.10   CMethaneSensor

The CMethaneSensor is responsible for the following functions (Figure B.17):

a. Extending CSensor
b. Reading methane level
c. Storing methane level.

Methane Elevation

Set allowable levels of methane based on a percentage (Figure B.18). Any reading above the critical level will trigger ventilation to go into an "on" state.

Methane Sensor State

This represents the sensor state based on the sensor reading (Figure B.19).

Methane Sensors Query Time

This represents the time required to make a sensor reading in seconds (Figure B.20).

CMethane:: **CMethaneState**
-m_state : METHANESENSORSTATE
+_init_(in name : <unspecified> = SENSORSTATENAME_METHANE)

FIGURE B.16   CMethaneState class

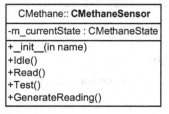

CMethane:: **CMethaneSensor**
-m_currentState : CMethaneState
+_init_(in name) +Idle() +Read() +Test() +GenerateReading()

FIGURE B.17   CMethaneSensor class

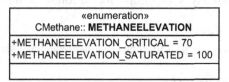

«enumeration» CMethane:: **METHANEELEVATION**
+METHANEELEVATION_CRITICAL = 70 +METHANEELEVATION_SATURATED = 100

FIGURE B.18   Methane level enumerations

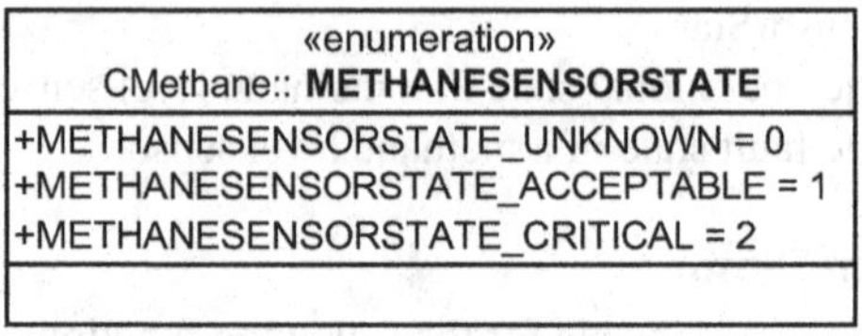

«enumeration»
CMethane:: **METHANESENSORSTATE**
+METHANESENSORSTATE_UNKNOWN = 0
+METHANESENSORSTATE_ACCEPTABLE = 1
+METHANESENSORSTATE_CRITICAL = 2

FIGURE B.19   Methane sensor state enumerations

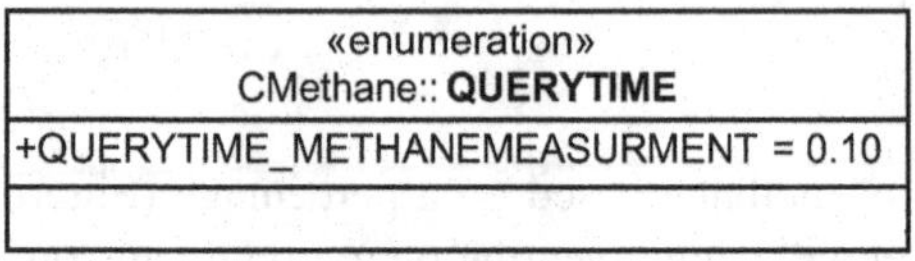

«enumeration»
CMethane:: **QUERYTIME**
+QUERYTIME_METHANEMEASURMENT = 0.10

FIGURE B.20 . Methane sensor reading time

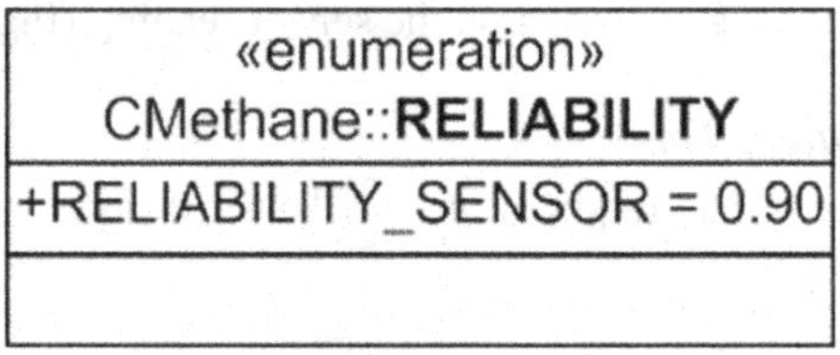

«enumeration»
CMethane::**RELIABILITY**
+RELIABILITY_SENSOR = 0.90

FIGURE B.21   Methane sensor reliability

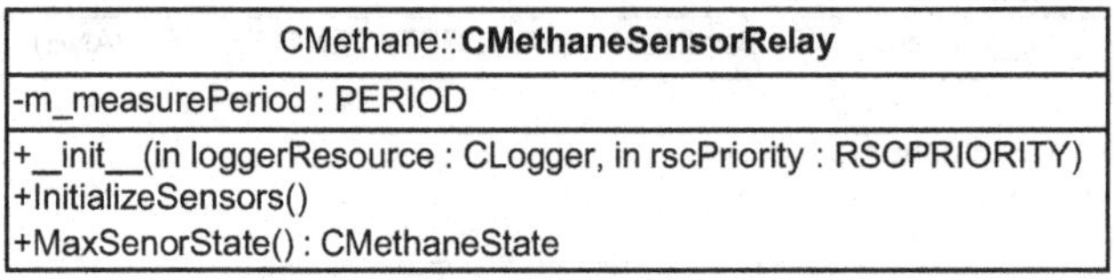

CMethane:: **CMethaneSensorRelay**
-m_measurePeriod : PERIOD
+_init_(in loggerResource : CLogger, in rscPriority : RSCPRIORITY)
+InitializeSensors()
+MaxSenorState() : CMethaneState

FIGURE B.22   CMethaneSensorRelay class

Methane Sensors Reliability

This represents the reliability of the sensor in the percentage of time correctly operating (Figure B.21).

B.3.2.11   CMethaneSensorRelay

The CMethaneSensorRelay is responsible for the following functions (Figure B.22):

1. Extending CSensorRelay
2. Providing process control for managing methane level sensors.

Number of Methane Sensors

This represents the number of sensors managed by the CMethaneSensorRelay (Figure B.23).

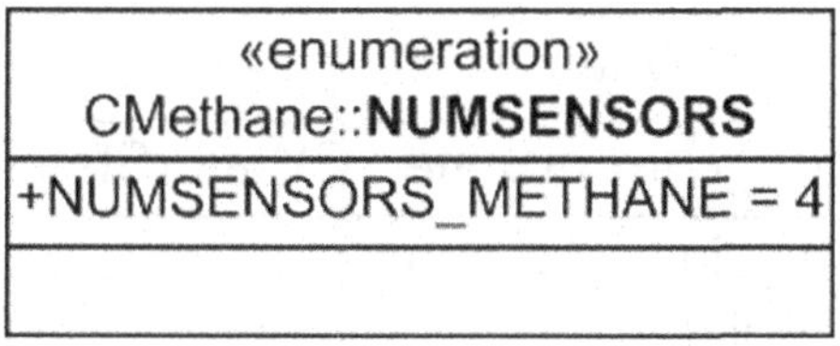

«enumeration»
CMethane::**NUMSENSORS**
+NUMSENSORS_METHANE = 4

FIGURE B.23   Number of methane sensors enumeration

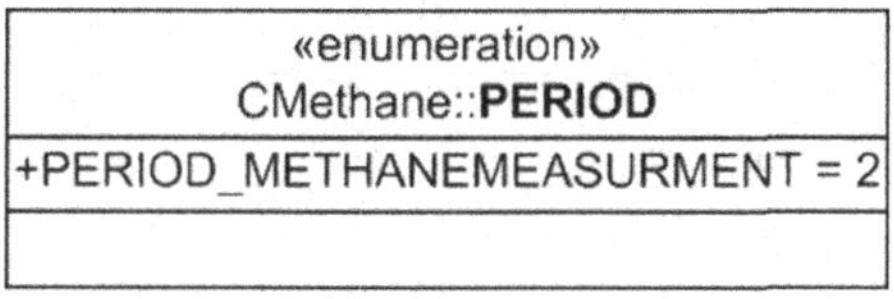

«enumeration»
CMethane::**PERIOD**
+PERIOD_METHANEMEASURMENT = 2

FIGURE B.24   Methane level reading period enumeration

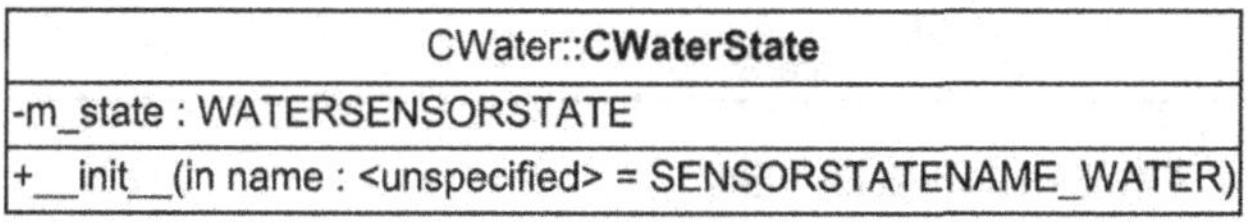

CWater::**CWaterState**
-m_state : WATERSENSORSTATE
+__init__(in name : <unspecified> = SENSORSTATENAME_WATER)

FIGURE B.25   CWaterState class

Methane Sensor Reading Period
This represents the time between successive sensor readings in seconds (Figure B.24).

B.3.2.12   CWaterState
The CWaterState is responsible for the following functions (Figure B.25):

1. Extending CSensorState
2. Maintaining the operational state of a water level sensor
3. Maintaining the fault state of a water level sensor.

B.3.2.13   CWaterSensor
The CWaterSensor is responsible for the following functions (Figure B.26):

1. Extending CSensor
2. Reading water level
3. Storing water level.

B.3.2.14   CWaterSensorRelay
The CWaterSensorRelay is responsible for the following functions (Figure B.27):

1. Extending CSensorRelay
2. Providing process control for managing water level sensors.

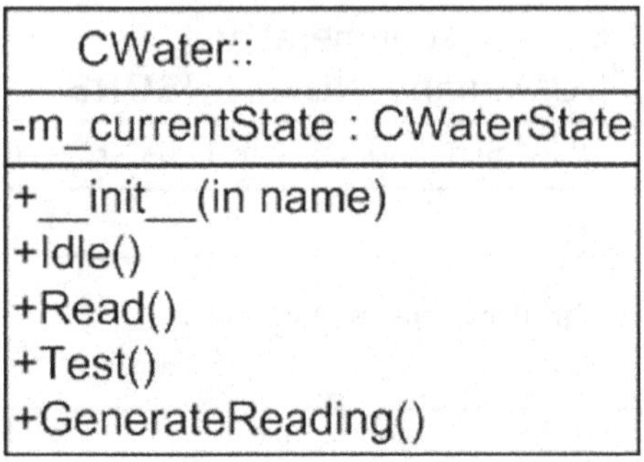

CWater::
-m_currentState : CWaterState
+__init__(in name) +Idle() +Read() +Test() +GenerateReading()

FIGURE B.26   CWaterSensor class

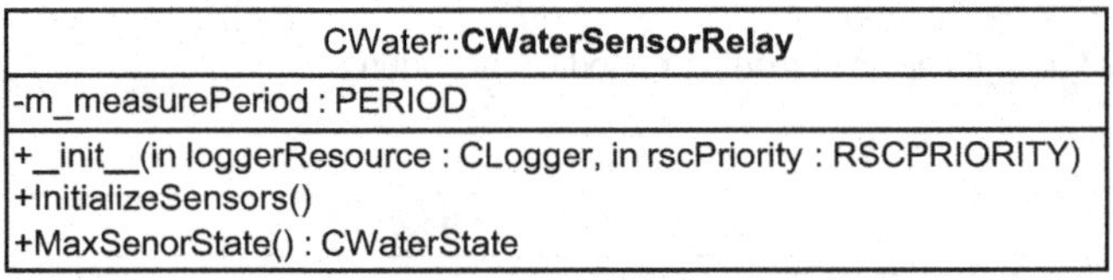

CWater::**CWaterSensorRelay**
-m_measurePeriod : PERIOD
+_init_(in loggerResource : CLogger, in rscPriority : RSCPRIORITY) +InitializeSensors() +MaxSenorState() : CWaterState

FIGURE B.27   CWaterSensorRelay class

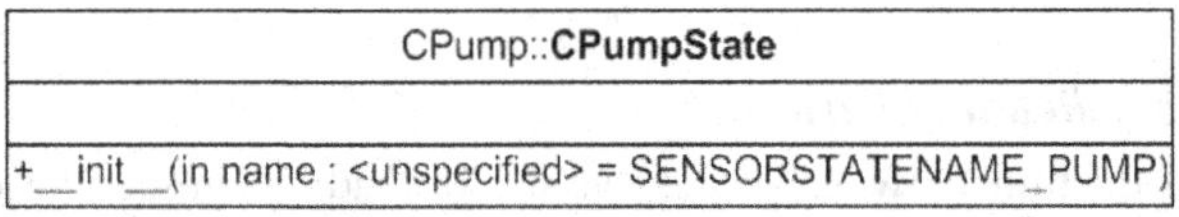

CPump::**CPumpState**
+__init__(in name : <unspecified> = SENSORSTATENAME_PUMP)

FIGURE B.28   CPumpState class

B.3.2.15 CPumpState

The CPumpState is responsible for the following functions (Figure B.28):

1. Extending CAbstractSensorState
2. Maintaining the operational state of a pump on/off sensor
3. Maintaining the fault state of a pump on/off sensor.

B.3.2.16 CPumpSensor

The CPumpSensor is responsible for the following functions (Figure B.29):

1. Extending CSensor
2. Reading the pump on/off state
3. Storing the pump on/off state.

B.3.2.17 CPumpSensorRelay

The CPumpSensorRelay is responsible for the following functions (Figure B.30):

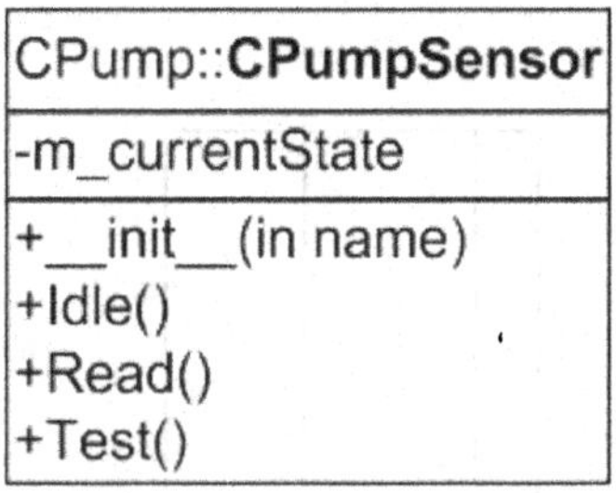

```
┌─────────────────────────┐
│ CPump::CPumpSensor      │
├─────────────────────────┤
│ -m_currentState         │
├─────────────────────────┤
│ +__init__(in name)      │
│ +Idle()                 │
│ +Read()                 │
│ +Test()                 │
└─────────────────────────┘
```

FIGURE B.29   CPumpSensor class

1. Extending CAbstractSensorRelay
2. Providing process control for managing pump on/off operation depending on current water level sensor readings.

B.3.2.18   CVentilationState

The CVentilationState is responsible for the following functions (Figure B.31):

1. Extending CAbstractSensorState
2. Maintaining the operational state of a ventilation fan sensor
3. Maintaining the fault state of a ventilation fan sensor.

B.3.2.19   CVentilationSensor

The CVentilationSensor is responsible for the following functions (Figure B.32):

1. Extending CSensor
2. Reading ventilation on/off state
3. Storing ventilation on/off state.

B.3.2.20   CVentilationSensorRelay

The CVentilationSensorRelay is responsible for the following functions (Figure B.33):

1. Extending CAbstractSensorRelay
2. Providing process control for managing pump on/off operation depending on current methane level sensor readings.

B.3.3   SEQUENCE DIAGRAM

The sequence diagram is shown in Figure B.34.

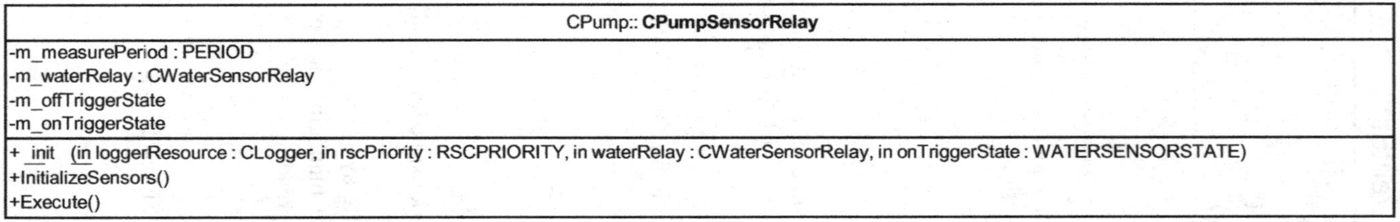

CPump:: **CPumpSensorRelay**
-m_measurePeriod : PERIOD -m_waterRelay : CWaterSensorRelay -m_offTriggerState -m_onTriggerState
+ init (in loggerResource : CLogger, in rscPriority : RSCPRIORITY, in waterRelay : CWaterSensorRelay, in onTriggerState : WATERSENSORSTATE) +InitializeSensors() +Execute()

FIGURE B.30 CPumpSensorRelay class

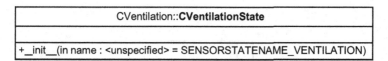

CVentilation::**CVentilationState**
+_init__(in name : <unspecified> = SENSORSTATENAME_VENTILATION)

FIGURE B.31   CVentilationState class

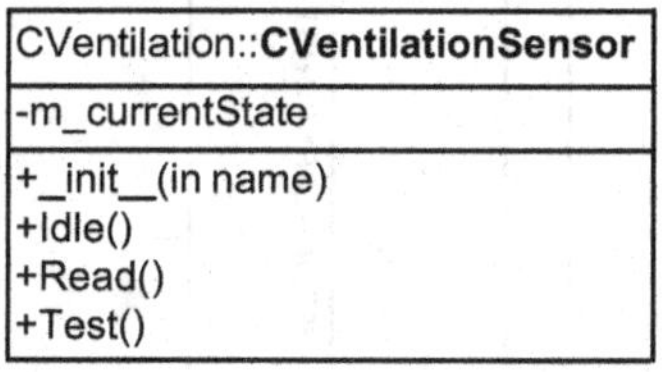

CVentilation::**CVentilationSensor**
-m_currentState
+_init__(in name) +Idle() +Read() +Test()

FIGURE B.32   CVentilationSensor class

CVentilation:: **CVentilationSensorRelay**
-m_measurePeriod : PERIOD -m_methaneRelay : CMethaneSensorRelay -m_offTriggerState -m_onTriggerState
+ init_ (in loggerResource : CLogger, in rscPriority : RSCPRIORITY, in methaneRelay : CMethaneSensorRelay, in onTriggerState : METHANESENSORSTATE) +InitializeSensors() +Execute()

FIGURE B.33 CVentilationSensorRelay class

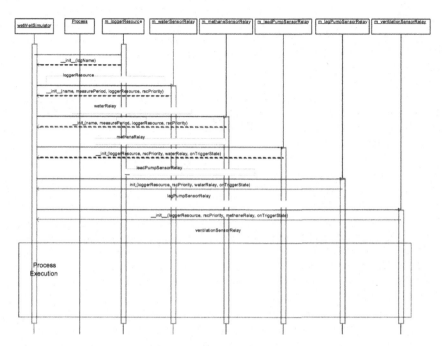

FIGURE B.34 Wet well controller (© 2007 by Taylor & Francis Group, LLC)

REFERENCES

1016-2009 – IEEE Standard for Information Technology—Systems Design—Software Design Descriptions

IEEE Recommended Practice for Software Requirements Specifications (IEEE Std. 830-1998).

Appendix C
Object Models for a Wastewater Pumping Station Wet Well Control System

Christopher M. Garrell

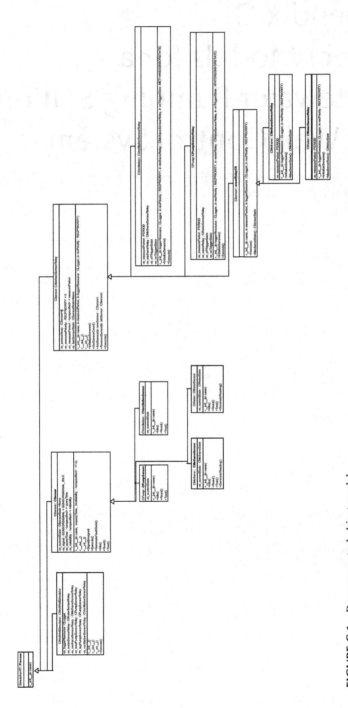

FIGURE C.1   Process control object model

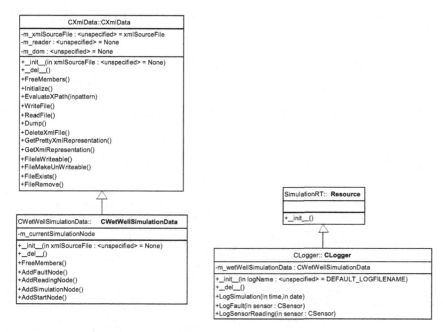

FIGURE C.2   Resource control object model

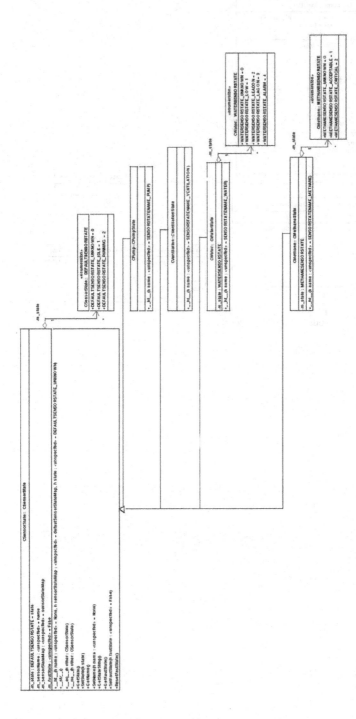

FIGURE C.3 Sensor state object model

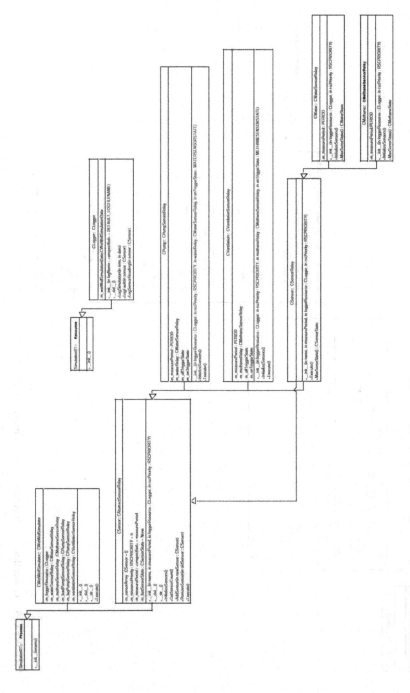

FIGURE C.4 General wet well control system object model

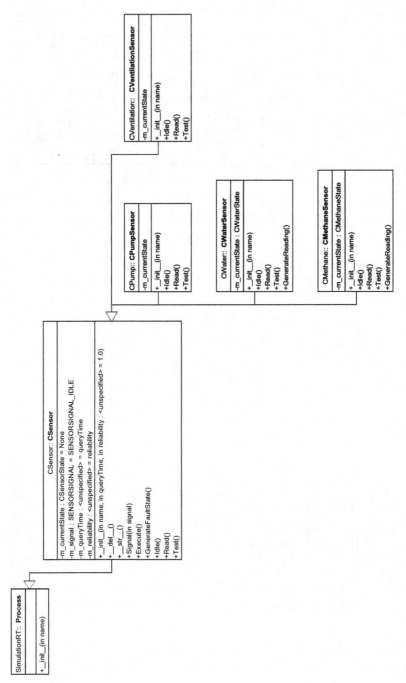

FIGURE C.5   General wet well control system object model

Appendix D
Unified Modeling Language

D.1 INTRODUCTION

Unified modeling language (UML) is a system of concepts and notation for abstracting and representing discrete systems, particularly but not exclusively object-oriented software systems. It is widely used as a common notation for describing software systems in publications and design models. Although widely misunderstood as merely a pictorial notation, UML is more importantly a set of modeling concepts that are more general than those contained in most programming languages and widely applicable to real-world discrete systems as well as software systems. UML is "unified" because it combines a number of competing but similar modeling languages that were previously fighting for domination, eventually superseding all or most of them in public usage; "modeling" because its primary intended use is to construct abstract models of discrete systems that can subsequently be converted into a concrete implementation, such as a programming language, database schema, or real-world organization; and a "language" because it comprises both internal structural concepts (metamodel) and an external representation (visual notation). UML supports an engineering-based design approach in which models of software are constructed to understand and organize a system and fix problems before software is written. Although modeling is universally used by engineers, architects, and economists, software personnel have often been reluctant to use modeling despite the large number of errors often found in completed software.

UML comprises a number of loosely connected facets, often called submodel types or (in reference to the visual notation) diagram types. UML is intended to be a general purpose modeling language; that is, it is applicable to all or at least most kinds of modeling problems. Unlike many academic modeling languages that emphasize elegance, a minimal set of basic concepts, and efficacy in theorem proving, UML is a pragmatic, "messy" modeling language with some redundant concepts and multiple ways of expressing them, similar to most mainline programming languages or natural languages. The various submodel types vary in elegance, level of abstraction, and usefulness. Because different submodel types were proposed and defined by different people as part of a consensus-based development process, they sometimes fit together a little awkwardly, like many major software applications.

UML is informed by object-oriented principles, although it can be used to model nonobject-oriented systems as well. It can be used at various levels of abstraction from very high-level system models to detailed models of programs.

D.2 DIAGRAM TYPES

In this description the term "diagram type" is used as a shorthand to represent a set of related concepts within a submodel type of UML and their visual notation. The most commonly used UML diagram types include class diagrams, use case diagrams, sequence diagrams, state machine diagrams, and activity diagrams.

D.2.1 CLASS DIAGRAM

The class diagram is by far the most widely used kind of UML diagram. It describes the concepts within a system and their relationships. It is used to organize the information within a system into classes, understand the behavior of the classes, and generate their data structure. Figure D.1 shows a (much simplified) example of a class diagram showing the kinds of recording requests within a digital video recorder (DVR) system attached to a satellite television receiver.

A class is the description of a system concept that can be represented by data structure, behavior, and relationships. For example, "RecordRequest" is a class representing an entry in the DVR list of television shows to be recorded. An individual instance of a class is an object. A class has attributes, each of which describes a data value held by each object that is an instance of the class. For example, a RecordRequest has a numerical priority indicating its urgency if the DVR disk becomes full. Each object has its own particular data values for the attributes of its class, and the values of each object may differ from other objects of the same class. The data values of

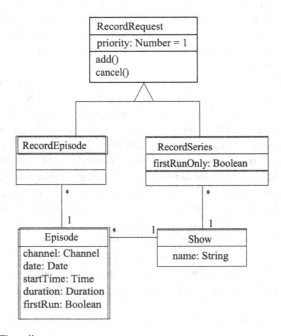

FIGURE D.1 Class diagram

a single object can vary over time. An attribute includes a name, a data type, and optional constraints on the values that objects of the class may hold. A class can also have operations, each of which describes a behavior that objects of the class can perform. Individual objects do not have distinct operations; all the objects of the same class have the same behavior, which can be implemented by shared code, similar machinery, printed documents, or other ways appropriate to the kind of system.

Classes are shown as rectangles with three sections, some of which may be optionally suppressed in a diagram. The top section shows the name of the class and the middle section shows a list of attributes in the form "name: type = initial value."

In the example, the default priority of a record request is 1. The bottom section shows a list of operations in the form "name (parameter-list): return-type." For example, a new record request can be added to the recording queue, and an existing request can be canceled. Various other indicators may be attached to attributes and operations.

There are two major kinds of relationships among classes: associations and generalizations. Associations are semantic relationships between pairs (or, in general, tuples) of objects. The value of an association is a set of object tuples. Associations are often implemented as pointers within objects (in C, C++, and Java) or as sets or tables of tuples of object identifiers (in SQL and other databases). Modeling associations is of great importance because they are the glue that holds a system together. A binary association, by far the most common kind, is shown as a line between two classes. For example, a show contains episodes. A number near an end of an association line indicates the multiplicity of the association in that direction, that is, how many instances of the class at that end may be associated with any one instance of the class at the other end. The value may be a single number or an integer range. An asterisk represents many instances, that is, zero or more without limit. For example, each show may include many episodes, but each episode belongs to exactly one show. A name can be given to the entire association and to both ends of it. In the example, there is a single unique association between each pair of classes, so the names have been omitted. Various other restrictions can be attached to the association ends.

Generalization is the relationship between a general class and a more specific variety of it. The general class (superclass) can be specialized to yield one of several specific classes (subclasses). This has several benefits. The data structure and behavior of the superclass can be inherited by each subclass without the need for repetition. Most importantly, an operation in a superclass can be defined without an implementation, with the understanding that an implementation will be eventually supplied by every subclass. This is called polymorphism, because the abstract operation takes on a different concrete form in each subclass. A class lacking implementations for one or more operations is an abstract class; it cannot be instantiated directly, but it can serve as a data type declaration for variables that can be bound to objects of any subclass. This means that operations can be written in terms of abstract classes, with the actual implementations dependent on the actual object types at run time. It also permits collections of objects of mixed subclasses to be freely manipulated; the operation implementation appropriate to each object in the collection is determined dynamically at run time.

Generalization is shown as a line between classes with a large triangle attached to the more general class. A group of common subclasses can be shown as a tree sharing a single triangle to the superclass. In the example, RecordRequest is a superclass of "RecordEpisode" and "RecordSeries." This indicates that a request for a recording may either indicate a single episode or all the episodes of a given series. Both RecordEpisode and RecordSeries inherit the attribute ("Priority") and operations ("Add" and "Cancel") from the superclass RecordRequest. The main advantage of inheritance is that subsequent changes to the superclass, such as the addition of new attributes or operations, are automatically incorporated in the subclasses.

From the example, we see that a series recording request can be indirectly related to a set of episodes. This might suggest that the implementation of the internal recording mechanism need only deal with episodes. The relationship between episodes and shows can be maintained within the request queue. Class diagrams permit this kind of analysis to be made quickly.

Several other kinds of relationships such as various kinds of dependencies and the derivation of certain attributes from others are less frequently used.

D.2.2 Use Case Diagram

A use case diagram shows the externally visible behavior of a system as it might be perceived by a user of the system. Such users can include humans, physical systems, and software systems. The diagram is used to factor and organize the high-level behavior of a system into meaningful units during early design, but detailed implementation is usually deferred to other, more specific models, such as activity diagrams or collaboration diagrams; Figure D.2 shows a use case diagram for a DVR (again, much simplified).

An actor represents a set of external users playing a similar role in interacting with the system being modeled. For example, actors that interact with the DVR are the viewer and the network. A use case represents a meaningful unit of behavior performed by the modeled system for the benefit of or at the request of a particular actor. For example, the viewer can request a recording, play an existing recording, and buy a pay-per-view show. (A real DVR would have a number of additional use cases that would make the example too large.) The network is involved in the "BuyShow" use case initiated by the user, and also participates in the "DownloadGuide" use case, which represents downloading the program guide periodically in the background without the involvement of the viewer. The granularity of a use case is somewhat subjective, but it should be large enough to be meaningful to the actor, rather than trying to capture the implementation of the system.

For example, it is not appropriate to include minor user interface actions such as paging through the program guide, selecting a show, and specifying the recording parameters as separate use cases; these are best modeled as details of the user interface that implement the use case. The purpose of enumerating use cases is to capture the externally visible behavior of a system at a high level.

The target system or class is shown as a rectangle containing a set of use cases, each shown as an ellipse containing the name of the use case. In the example, the

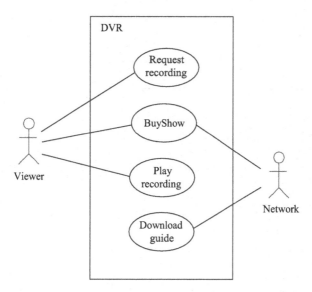

FIGURE D.2   Use case diagram

main focus is the DVR, its use cases, and the objects that communicate with it directly. An actor is shown as a stick person outside the bounds of the system rectangle. In the example, the viewer and network are actors with respect to the DVR. A line connecting an actor to a use case shows the association between the actor and the use case in which it participates. Each actor can participate in multiple use cases and each use case can interact with multiple actors. Details of the actors themselves are not included in the use case diagram; the focus of the use case diagram is the behavior of the target class, in this case the DVR. If some of the actors are also systems, they might have their own use case diagrams.

Various relationships among use cases can be shown, such as generalization and the inclusion of one use case as a part of another.

The use case diagram does not contain a great deal of detailed information. It is primarily of use in the early stages of design to organize the system requirements, particularly from a user viewpoint. The behavior of a use case can be documented using sequence diagrams, activity diagrams, or, in most cases, simple text descriptions.

D.2.3   Sequence Diagram

A sequence diagram shows an interaction among two or more participants as a set of time-sequenced messages among the participants. The main focus is the relative sequence of messages, rather than the exact timing. It is used to identify and describe typical as well as exceptional interactions to ensure that system behavior is accurate and complete. Sequence diagrams are particularly useful for interactions among more than two participants, because two-party conversations are easily shown as simple text dialogs alternating between the two parties. Figure D.3 shows an example of a sequence diagram for viewing a live show using the DVR.

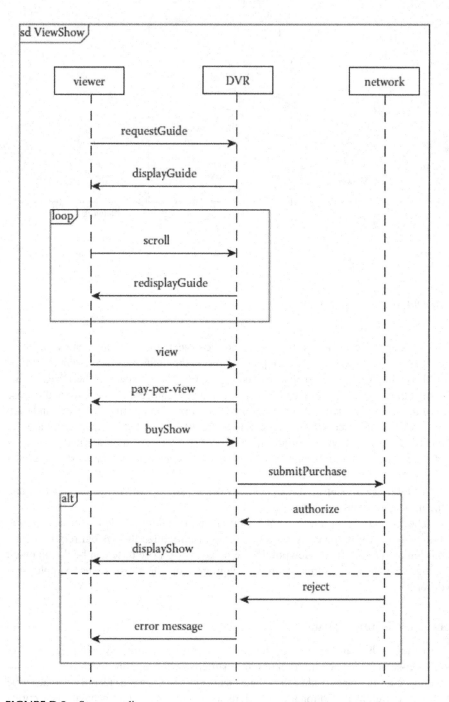

FIGURE D.3   Sequence diagram

A sequence diagram involves two or more roles, that is, objects that participate in the interaction. Their names (and optionally types) are shown in small boxes at the top of the diagram, with vertical lines (lifelines) showing their lifetime in the system. The example shows three roles: the viewer, the DVR, and the satellite network. In this example, all three objects exist for the duration of the interaction; therefore, their lifelines traverse the entire diagram.

A message is a single one-way communication from one object to another. A message is shown as a horizontal line between lifelines with an arrowhead attached to the receiving lifeline. The name of the message is placed near the line. In a more detailed diagram, the parameters of the messages can be shown. The example begins when the viewer requests to see the program guide, which the DVR then displays. The viewer may ask to scroll the guide, which the DVR redisplays. This subsequence is placed inside a "loop" box to show that it may be repeated. The viewer then requests to view a show. In this example, the show happens to be a pay-per-view show, so the DVR asks the viewer whether the viewer wants to buy the show. The user does buy the show, so the DVR submits the request to the network. The example shows two possible outcomes in the two sections of the alt (for "alternative") box: If the purchase is successful, the network sends an authorization and the DVR displays the shows; if the purchase fails (maybe the viewer's credit limit has been exceeded), an error message is displayed. Although a sequence diagram can display closely related alternatives, it is not intended to show all possible interactions. It is meant to show typical interaction sequences to help users understand what a system will do and to help developers think about possible interactions among features. Instead of trying to show everything on a single sequence diagram, a model should include multiple sequence diagrams to represent variant interactions. It is particularly important to include at least one sequence diagram for every possible error condition or exception that can occur.

D.2.4 STATE MACHINE DIAGRAM

A state machine diagram shows the states that an object (often a system) can assume and the transitions that move the object among states in response to various events that the object may perceive. It is used to identify and structure the control aspect of system behavior. Simple state machines have been important in electrical engineering and computing for many decades. UML uses an enhanced version of state machines based on the work of David Harel, which allows grouping of states into higher-level states. This is a kind of generalization of states that avoids the repetition of the same transition in many low-level states. State machine diagrams are useful for traditional control applications as well as for understanding the life history of objects.

Figure D.4 shows an example of a state machine diagram showing the life history of a recording request in the DVR. The same state machine is applicable to every recording request. Many recording requests can exist simultaneously, and each has its own distinct state.

A state is shown as a rounded rectangle containing the name of the state. For example, the "Queued" state means that a recording request has been created, but the

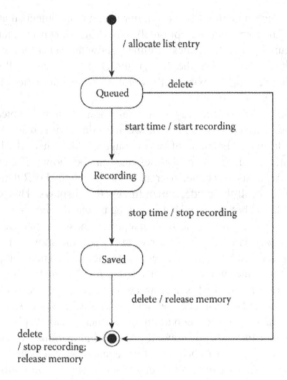

FIGURE D.4   State machine diagram

start time of the show has not yet arrived. A transition is shown as an arrow from a source state to a target state. The transition is labeled with the name of an event that may occur during the source state and, optionally, with a slash (/) character followed by an action. If an event occurs when an object is in a state that is the source state for a transition for that event, the action on the transition is performed and the object assumes the target state specified by the transition. For example, if the start time of a show arrives while a recording request is in the Queued state, the DVR starts recording, and the recording request assumes the "Recording" state. That state in turn lasts until the stop time of the show arrives, after which the recording ceases and the recording request assumes the "Saved" state.

The creation of an object is shown by a small black disk. The disk is connected by a transition to the initial state of the object; the transition may also specify an initial action. For example, when a recording request is created, a list item is allocated for it and it assumes the Queued state.

The destruction of an object is shown by a small bull's–eye. If a transition causes the object to reach this state, any action on the transition is performed and the object is destroyed. For example, if the "Delete" event occurs while a recording request is in the recording state, the memory for the recording is deallocated and the request is destroyed; if it occurs in the "Queued" state, no recording has yet occurred, so the request is just destroyed.

The full version of state diagrams supports concurrent states. The state of an object can have two or more separate parts, each of which progresses independently. For example, a more elaborate version of the DVR state machine diagram would show the ability both to record a show and to play an earlier portion of the recording simultaneously. Although the concurrent states are mostly independent, they can have some dependencies. For example, if the viewer fast-forwards the playback of the recording so that it catches up with the live recording, the viewer cannot jump forward any further.

Substates are also supported by nesting substates within superstates. For example, the Delete event could be attached to a single transition on a high-level super-state containing multiple detailed substates. If the event occurred during any of the detailed substates, the high-level transition would take precedence. The ability to apply a single transition to many nested substates is particularly useful for exception conditions and forced exits, because it becomes unnecessary to attach the error handling to each nested substate explicitly.

D.2.5 ACTIVITY DIAGRAM

An activity diagram shows the sequencing of actions within an overall process. In general it includes concurrent processes as well as sequential processes, so it has the form of a directed graph rather than a simple chain. This diagram is used to understand the control and data dependencies among the various pieces of detailed behavior that compose a high-level system behavior, to ensure that the overall behavior is adequately covered and that detailed behaviors are performed in the correct orders.

The sample activity diagram in Figure D.5 shows the process of changing a channel on the DVR. An activity diagram can be divided into multiple columns, each showing the actions performed by one of the participants. The participants in this example are the viewer, the DVR program guide, and the satellite receiver subsystem. An action is shown as a rounded rectangle. This is the same symbol as a state in a state diagram because an action is a state of a process. A decision is shown as a diamond with two or more outgoing arrows, each labeled by a condition. An arrow shows a temporal dependency between two actions; the first must be completed to enable the second. Synchronization is shown by a heavy bar with two or more inputs or outputs; a synchronization bar initiates or terminates concurrent actions.

In the example, initially the viewer selects a channel. This triggers the program guide subsystem to determine if the viewer subscribes to the channel and if it is currently broadcasting. If the channel is not available, the viewer receives an error message. If the channel is available, the receiver must perform three concurrent actions to display the channel, as shown by the synchronization bar followed by arrows to three concurrent actions. The decision performed by the program guide enables the three concurrent actions: tuning the receiver, selecting the correct satellite input, and setting the television format to low or high definition. The initiation of concurrent activity is called a fork. Once concurrent actions are enabled, they proceed independently until a subsequent synchronization bar joins them back together. When all three actions are complete, the synchronization bar on the bottom is triggered, which

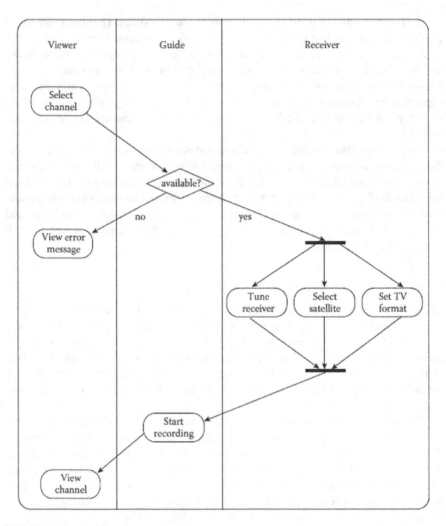

FIGURE D.5   Activity diagram

triggers the guide unit to start recording the current channel. The synchronization of concurrent activity is called a join. Finally the user is allowed to view the channel.

In the complete version of activity diagrams, conditional and iterative behavior may be represented, as well as more complicated kinds of synchronization.

Activity diagrams can be used to plan implementations of operations. They are also useful for modeling business processes.

D.3   OTHER DIAGRAM TYPES

UML has several other diagram types that are not illustrated in this entry. See the references for more details.

D.3.1 INTERNAL STRUCTURE DIAGRAM

An internal structure diagram is a variant on the class diagram that shows the internal structure of an object in terms of its constituent parts and their relationships to each other and to ports connecting the object to the external world. This diagram type is used for hierarchical design, in which an external view of a class at one level is expanded into a particular realization or internal implementation at the next lower level. This is a more recent yet important addition to UML that bridges the gap between hierarchical specification and implementation more effectively than simple class diagrams, although widespread adoption has been slow.

D.3.2 COLLABORATION DIAGRAM

A collaboration diagram shows a collaboration. A collaboration is a contextual relationship among a set of objects, each of which plays a particular role within the collaboration. For example, a dinner party is a collaboration in which roles might include host, guest, and server. A collaboration has structure and behavior. This diagram is used to understand the specific behavior of objects in a contextual relationship, as opposed to behavior inherent to an object in any context because of its class.

D.3.3 COMMUNICATION DIAGRAM

A communication diagram shows a sequence of messages imposed on the data structure that implements the system. It combines aspects of sequence diagrams and internal structure diagrams in a rich form that is, however, more difficult to perceive. It is useful for designing algorithms and code provided excessive low-level detail is avoided.

D.3.4 DEPLOYMENT DIAGRAM

A deployment diagram shows the physical implementation elements of a system, including the physical execution nodes and the software artifacts stored on them. This diagram type lacks the detail of some other diagram types and is not very concise; therefore, it is not among the most successful UML diagram types.

D.3.5 PACKAGE DIAGRAM

A package diagram shows the organization of a software system into design and compilation packages, including nested packages and their dependencies. It is used to organize a model into working units and to understand the effects of changes on the overall system.

D.4 PROFILES

UML is a large, general purpose modeling language, but frequently the full power of UML is not needed in a particular usage. At the same time, sometimes a model builder wants to adopt more stringent assumptions than those made by the full UML, which must accommodate a wide range of uses. A UML profile is intended for this kind of usage. For example, UML in general assumes full concurrency, but a profile for a sequential application domain might remove this assumption to simplify modeling within that domain.

A profile is a named, specified subset of UML together with a limited set of additions and constraints that fit the profile for a particular area of usage. Profiles are often constructed and supported by standardization workgroups responsible for a particular application domain. They provide a way to adapt UML to a particular application area rather than constructing a new complete modeling language.

Because profiles are layered on standard UML using a limited set of extension mechanisms, they can be constructed and maintained by general purpose UML modeling tools.

A lot of modeling energy has gone into the construction of profiles. These include profiles for various programming languages, database schemas, system engineering, real-time systems, and a wide variety of application domains.

D.5   USAGE

UML can be used in many different ways for different purposes. One advantage of using UML rather than specialized languages for various domains is that most real-world applications include subsystems of many different kinds, and most specialized languages are somewhat restrictive in their scope, leading to problems in combining models built using different languages. A number of commercial and open-source tools support editing of UML models; the tools vary widely in completeness and other features, such as support for code generation.

D.5.1   PROGRAMMING DESIGN

UML can be used to perform the high-level design of software systems by first capturing requirements using class diagrams, use case diagrams, and sequence diagrams. The class diagrams can be elaborated to design classes that will be implemented in programming languages. Package diagrams can help to organize the structure of the program units.

D.5.2   PROGRAMMING DETAIL

Detailed class diagrams can be used to generate class declarations in languages such as C++ and Java. A number of commercial software design tools can perform this kind of code generation. Specification of algorithms can be done using activity diagrams and state machine diagrams, although these can sometimes be as expansive as the final code. Code generation using patterns can produce a significant amplification of detail and may be the most productive use of modeling to generate code.

D.5.3   DATABASE DESIGN

Class diagrams can model the information to be represented within a database. A number of UML extensions have added database features (such as keys and indexes) to basic UML. The database community has been somewhat wary of UML, however, often preferring to use older modeling languages explicitly designed for databases, but often lacking in expressiveness. One advantage of using UML is that both database schemas and programming language class declarations can be generated from the same information, reducing the danger of inconsistency.

D.5.4 BUSINESS PROCESS DESIGN

Activity diagrams are good for modeling business processes. Class diagrams can model the data implicit in business processes, usually with more power than traditional business modeling languages.

D.5.5 EMBEDDED CONTROL

State machine diagrams can represent the control aspects of a real-time system, but class diagrams and internal structure diagrams can also represent the data structures that the system uses.

D.5.6 APPLICATION REQUIREMENTS

Use case diagrams can capture the high-level functionality of a system. Class diagrams can capture the information in a particular application domain. These can then be elaborated to produce a more detailed design model.

D.6 UML HISTORY

Modeling languages for various purposes date to at least the early 1970s. Although some were widely used in specific application domains, such as telephone switching systems, none achieved widespread general usage. In the early 1990s a handful of object-oriented modeling languages developed for general use gained some popularity with no single language being dominant. These languages incorporated ideas from various earlier modeling languages but added an object-oriented spin. The first version of UML was created in 1995 by James Rumbaugh and Grady Booch by uniting their object modeling technique (OMT) and Booch methods. The bulk of that union still forms the core of UML, including the class diagram, the state machine diagram (including Harel's concepts), the sequence diagram (subsequently much extended with ideas from message sequence charts), and the use case diagram of Jacobson. Subsequent contributions by Ivar Jacobson and a team of submitters from a number of companies resulted in standardization of UML 1.0 by the Object Management Group (OMG) in 1997. The rights and trademarks to UML were then transferred to the OMG. OMG committees generated several minor updates to UML and a major update to UML 2.0 in 2004 that added profiles and several new diagram types, including internal structure diagrams and collaboration diagrams. OMG continues to maintain the standard. The complete standard and information on standardization activities can be found on their Web site at www.omg.org by following links to UML.

D.7   ASSESSMENT

UML has achieved its goal of uniting and replacing the multitude of object-oriented modeling languages that preceded it. It is frequently used in the literature to document software designs. It has been less successful in penetrating the practices of average programmers. Partly this is due to the misperception that UML is primarily just a notation, whereas its main purpose is to supply a common set of abstract concepts useful for modeling most kinds of discrete systems. Another factor, however, is that the various parts of UML are widely uneven in quality, approachability, and detail. Some parts, such as class diagrams, are widely used, whereas others, such as deployment diagrams, fail to live up to their promise. One problem is that the committee process of an organization such as the OMG encourages messy compromises, redundancy, and uneven work at the expense of cohesion. There is also an unresolved tension in what the modeling language should be, from an elegant and minimal theoretical basis at one extreme to a "programming through pictures" tool at the other extreme, with the more useful abstract modeling language somewhere in the middle. The tension about what UML should be has worked itself out differently in different submodel types and even in various parts of the same submodel type, which can lead to confusion and uneven coverage of modeling. To its credit, UML is a universal language with features to support a wide range of systems. Such languages, including practical programming languages, must be large and messy to some extent if they are to be used by a wide range of users on a wide range of problems. UML would undoubtedly benefit from a serious trimming of redundant and less successful concepts, but whether that can occur within the limits of a community process is doubtful. UML will probably remain a useful but far-from-perfect tool, but the window for a more coherent modeling language may well be closed.

FURTHER READING

See the OMG Web site for a detailed specification of UML (UML Specification), including the latest additions to the standard as well as ongoing standardization activity. The UML Reference Manual (Rumbaugh, Jacobson, & Booch, 2005) provides the most coherent detailed reference to UML, more highly organized than the highly detailed standard documents. The UML User Guide (Booch, Rumbaugh, & Jacobson., 2005) provides an introduction to UML organized by diagram type. Fowler's *UML Distilled* (Fowler, 2004) is a brief and highly selective introduction to the parts of UML that its author feels are most important, including some overall suggestions on a recommended design approach. Many other books describe UML either in whole or focused on particular usages.

REFERENCES

Booch, G., Rumbaugh, J., & Jacobson, I. (2005). *The Unified Modeling Language User Guide.* 2nd ed. Addison–Wesley, Boston, MA.

Fowler, M. (2004). *UML Distilled: A Brief Guide to the Standard Object Modeling Language.* 3rd ed. Addison–Wesley, Boston, MA.

Harel, D. (1987). Statecharts: A visual formalism for complex systems. *Science of computer programming. 8.3:* 231–274.

Rumbaugh, J., Jacobson, I., & Booch, G. (2005). *The Unified Modeling Language Reference Manual. 2nd ed.* Addison–Wesley, Boston, MA.

UML Specification. www.uml.org (accessed April, 2022).

Index

Note: Page numbers in *italics* refer to figures; page numbers in **bold** refer to tables.

Printed in the United States
by Baker & Taylor Publisher Services